Manual editor
Robert M. Santcci

Series editors
Peter Werwath
Cecilia Cassidy

Contributors
Kent Buhl
William Duncan
Gene Ruckle

Copy editor
Jude Cashman

Single Family Selective Rehabilitation

- for Single Family Construction Managers
- Production Step-by-Step
- Model Policies & Procedures
- Forms and Documents

THE ENTERPRISE FOUNDATION

VNR VAN NOSTRAND REINHOLD
New York

Softcover reprint of the hardcover 1st edition 1991
Library of Congress Catalog Card Number 91-12555
ISBN 978-1-4615-9728-5 ISBN 978-1-4615-9726-1 (eBook)
DOI 10.1007/978-1-4615-9726-1

Manufactured in the United States of America

Published by Van Nostrand Reinhold
115 Fifth Avenue
New York, New York 10003

Chapman and Hall
2-6 Boundary Row
London, SE 1 8HN

Thomas Nelson Australia
102 Dodds Street
South Melbourne 3205
Victoria, Australia

Nelson Canada
1120 Birchmount Road
Scarborough, Ontario M1K 5G4, Canada

16 15 14 13 12 11 10 9 8 7 6 5 4 3 2 1

Library of Congress Cataloging-in-Publication Data

Single family selective rehabilitation / manual editor, Robert M. Santucci ; contributors, William Duncan, Gene Ruckle, Kent Buhl.
p. cm.—(Housing production manual ; #1)
"For rehab program managers; production step-by-step; model policies & procedures; forms and documents. "
Includes index.
ISBN 978-1-4615-9728-5
1. Housing rehabilitation—United States. 2. Housing, Single family—United States—Conservation and Restoration. 3. Poor- -Housing—United States. I. Santucci, Robert M. II. Duncan, William, 1949- . III. Ruckle, Gene. IV. Buhl, Kent. V. Enterprise Foundation . Rehab Work Group. VI. Series.
HD7293.S553 1991
363.5'8—dc20 91-12555
CIP

TABLE OF CONTENTS

EXAMPLES

ADDENDA

Foreword

The procedures in this manual distill six years of learning about the most efficient acquisition and construction methods in this country for high volume production of housing affordable to low-income people.

The people in the Rehab Work Group at The Enterprise Foundation have gathered this information first-hand and conveyed it in training workshops on production management throughout the country. Over 2000 participants in the workshops have included rehab specialists from city housing departments, construction managers of single and multifamily rehab projects, program managers from both nonprofit housing development organizations and city rehab departments, as well as students in colleges and universities. They are becoming the backbone of a new cadre of effective producers of affordable housing.

In the 30 cities and several rural areas where we have both learned and applied these production management techniques, they have resulted in savings of 5, 10, even 25 percent of the cost of housing development.

The Enterprise Foundation is a national, nonprofit organization that my wife Patty and I launched in 1982. The Foundation's mission is to see that all very low-income Americans have the opportunity for fit and affordable housing within a generation, and work their way up and out of poverty into the mainstream of American life.

Toward this end, the Foundation works with local nonprofit housing organizations to reduce construction costs, find low-rate financing, and increase their housing production and preservation activities. This has involved the renovation or construction of nearly 14,000 homes and apartments. In the process, the Enterprise Rehab Work Group has become, perhaps, foremost in the United States in researching, applying, and training others in low-cost construction and rehab techniques.

In addition to making available to housing professionals what we know as efficient production management systems, the publication of this book rewards The Enterprise Foundation in other ways:

- The proceeds will help us advance our work with local nonprofit housing groups
- The research supported by this project has already made our work more effective and will continue to do so.
- Our work will allow others to share the knowledge we have helped develop.
- And finally, we hope that widespread use of these techniques will lead to gradual change in the whole business of housing repairs and renovation—making ownership of housing more affordable for low and moderate income families.

Reduction in the cost of housing is important if every American is to have a decent home. Americans with low and moderate incomes must rely on preservation and rehabilitation of our aging, existing housing stock, and very low cost new construction.

The housing industry in facing this challenge is working to change outmoded regulations, identify ways to cut costs without sacrificing quality, and increase efficiency—by adopting advanced management techniques that offer practical solutions with lower overhead. We hope this manual contributes toward that end.

This book was only possible because of the resourceful work and strong backing of Peter Werwath, who has served as Director of the Foundation's Rehabilitation Work Group, and his staff—Cecilia Cassidy, Gene Ruckle, Bob Santucci and Bill Duncan, who is now Director of this energetic group. Gene Ruckle deserves special credit for directing important phases of this research project. This team has done the work in the field, and learned and taught the lessons the book provides.

James W. Rouse
Chairman
The Enterprise Foundation

Introduction

Single-Family Selective Rehabilitation (Housing Production Manual #1) has been written by the Enterprise Foundation's Rehab Work Group (RWG), a nonprofit organization that has helped to create construction management systems for nonprofit groups across the United States. It presents an idealized template for planning, running, and staffing an efficient housing rehab program, and can serve either as an aid to starting a new program or as a tool for improving an existing one.

This manual and the other two in RWG's series: **Multifamily Selective Rehabilitation (Housing Production Manual #2)** and **Substantial Rehabilitation/ New Construction (Housing Production Manual #3)**, represent the most current, detailed compilations of the efficient, accepted production management models available in low-income housing rehabilitation. The techniques that we recommend in the manuals are modeled after the most pervasive and effective strategies used in high-production programs.

The term **production management**, the core concept of these manuals, focuses on the problems of design, budgeting, and the physical rehabilitation of housing. Other areas of rehabilitation management, such as financing, marketing, intake, or underwriting processes, are not covered here, except when they intersect with the actual production process.

The manuals are designed to be process oriented; that is, they are laid out in a step-by-step progression that is intended to travel from the very beginning stages of program planning all the way through to post-construction and final performance evaluation.

The manuals are not designed to be slavishly followed as a rigid system. Rather, they are clearly meant to be adapted to the unique needs of a given locale and operation system. While improvised approaches are sure to waste time and money, poorly fitting adaptations from one rehab sector to another can totally thwart production. For example, using a 14-page architect's contract on a $5,000 rehab job will add unnecessary paperwork, staff time, and costs.

To expedite matters, we have included all of the major pieces of the process of rehabilitation, such as: the forms and model documents, ideas for gathering data on prospective projects, suggestions for creating specifications and work write-ups, systems to track budgeting and filing, ideas for finding contractors and/or subcontractors, practical suggestions for carrying out the bidding process, and techniques for ensuring the successful completion of projects.

You will find that the manuals dealing with selective rehab emphasize automation. Computers and computer software, such as Specmaster, the Enterprise Foundation's proprietary construction software, have quickly established themselves as indispensable aids in the process of creating specifications, producing cost estimates, creating bid documents, and creating work write-ups.

Your main goal and ours is efficient and, hopefully, increased production. Getting a standardized process worked out is just one means to that end. After reasonable goals have been set, a process chosen, and financing raised, innovation and entrepreneurship are still important. The task then is rejecting, accepting, or simplifying the details of process to achieve more and better results.

◙ A Note About Rehabilitation Management Systems

There are basically two distinct types of rehabilitation management systems currently being used in the construction industry in the U.S.: substantial rehab and selective rehab. While gut rehabilitation replaces complete building systems, selective rehab only replaces what is broken, about to break down, or missing on a room-by-room, unit-by-unit basis.

Gut rehabs essentially follow the traditional design and construction management systems of the U.S. building industry. They account for approximately 50,000 subsidized, low-cost units per year in the U.S. These methods rely on blueprints and general contractors. Units undergoing substantial rehab must be vacant during the rehab process. Standard industry practices are normally altered only by the peculiarities of site acquisition and the requirements of multisource public financing in relation to nonprofit housing production. Development costs average more than $50,000 per unit and construction costs, $30,000 per unit.

Selective rehab, on the other hand, is a more pervasive and economical rehab system that is particularly suited to small-scale and occupied rehabilitation projects. Selective rehab amounts to approximately 300,000 units per year in the U.S., at an average cost of less than $15,000 per unit. It is most pervasive in local rehab programs funded by community development block grant, rental rehabilitation, and Farmer's Home Administration programs, as well as local nonprofit efforts.

The methods used in selective rehab have emerged primarily out of the unstructured practices of the housing maintenance, repair, and remodeling industries as adapted by those federally funded and nonprofit programs. In contrast to gut rehab, this method often allows units to remain occupied during the process, although this is not economically feasible when more than 40 to 50 percent of components must be replaced.

◙ Computerization

The computer is a tool that is essential to efficient operation of a single-family selective rehab program. Computer programs are required for estimating, specification writing, bidding, job tracking, and accounting in this manual. While the best manual systems can compete with the less efficient, fully automated offices, the better automated programs can consistently renovate between 30 and 115 units per year per staff person. Computerization, hardware, software, and necessary staff training may cost $10,000 to $25,000, but usually pays back within the first year of high volume operation in reduced per unit costs.

Housing Types

There are two main housing types that have major implications on choices of production systems:

1. **Single-Family Buildings** These are defined as detached, semi-detached (e.g., duplex), or row houses in which common areas, that is, areas shared by two or more units, are minimal or even nonexistent. Legal title to the property is usually vested in individual units. This housing type is most often owned by individuals, but sometimes it is owned by landlords, such as the scattered-site homes in the South or the row houses in Eastern cities.

2. **Multifamily Housing** In the low-income housing field, these units are usually rentals and are owned privately, publicly, or cooperatively. Common areas are extensive, and it is generally not critical to isolate the design, budget, and construction of individual dwelling units.

Contracting Modes

Throughout these three process manuals, there are three basic ways of getting construction work done. They are defined as follows:

1. **Small-Scale General Contractors** These companies generally have two to five employees, and they specialize in rehabilitation or custom homebuilding. We believe the use of these contractors is the most appropriate choice for selective rehab, gut rehab, or new construction of single-family units, unless they are in a very large, scattered-site project. Contractors usually specialize in one of these three construction methods. Such contractors will take total responsibility for the construction work, and will hire and pay all necessary subcontractors.

2. **Construction Management** In this system, a landlord, developer, or a public or nonprofit intermediary agency employs a construction manager or management firm, which makes written work specifications, bids out the work to subcontractors and monitors quality of work and payments. In this instance, no one is legally liable for completion of the job, except the owner in some cases, although subcontractors and the construction

manager are liable for their specified jobs. This method of operation is used most frequently and appropriately for multifamily selective rehab, in our experience.

3. **Large-Scale General Contractors** Fitting the prevailing pattern of the U.S. architect/general contractor system, large-scale general contractors are clearly appropriate for larger gut rehab and new construction projects, such as those exceeding 10 units. In some cases, they may be "vest pocket contractors"; that is, contractors mainly employing subcontractors, or, alternately, they may use large numbers of their own employees. These contractors are defined as firms that can handle contracts of $500,000 or more.

◙ The Delivery Systems

Using the variables described above -- the kinds of rehab, the types of housing, and the types of contractors, rehabilitation of low-income housing can rationally be organized into the basic production models described in our three manuals and summarized in the following table. Obviously, these characterizations are not hard and fast, and some highly efficient activity does not follow this scheme.

Again, we emphasize that final choices on program design must be made at the local level. For example, local rehabbers may find the construction-management model works better for gut rehab, because of an absence of experienced general contractor, or because of a desire to integrate job training crews.

Types of Production Model	Preferred Design Method	Preferred Budget Method	Preferred Execution Method
Single-Family Selective Rehab	Written itemized specifications -- locational. Some graphic details.	By work item -- locational.	Small-scale general contractor.
Multifamily Selective Rehab	Written itemized specifications -- locational.	By subcontract (i.e., trades and line item).	Construction manager/sub-contractors.
Gut Rehab/New Construction	Blueprints.	By conven-tional building trades and line item.	General contractor.

◙ How to Use This Manual

Before you begin to use **Single-Family Selective Rehabilitation**, it is important that you understand how this manual has been designed. First of all, it is designed to use general contractors, rather than the construction management mode. Second, we have assumed you will be doing an annual volume of approximately 50 units a year.

Here is how we have broken down the program process from initial set-up through completion of the project:

Stage — The five basic stages presented in this manual are **Program Planning, Design and Feasibility, Bidding, Construction** and **Post-Construction**. Each stage represents a distinct component of the rehab process with many steps within it. Generally, one stage should be completed before moving on to the next.

Step "Set Up Contractor Recruitment and Evaluating System" is an example of a step in the manual. Steps are a collection of tasks that should be done consecutively. Generally, steps have been presented in a logical sequence, but the importance of following the exact order varies. Many of the steps are accompanied by model forms and sample documents. Longer documents related to a step have been included in the Addenda in the back of the manual and are referenced in the text.

Task Tasks are pieces of work that can generally be assigned to a single person, such as, "Conduct the Contractor recruitment meeting," Task 12.4.

Job Title To help you in program planning and staffing, we have assigned one or more of a small, idealized set of job titles or "departments" to each task. We assume that in a small program, one staff person might perform more than one "job," even though his or her title may not match the assigned title in this manual. In very large programs, layers of management and review would be necessary that are not presented here, and they might have more specialized technicians as well.

The job titles and departments we have selected to use in this manual are:

- Accounting Staff
- Board of Directors
- Executive Director
- Loan Officer
- Office Staff
- Program Manager
- Program Planner
- Rehab Specialist
- Senior Rehab Specialist
- Specmaster Clerk

(See Example 1, "Idealized Staffing Pattern for Single-Family Selective Rehabilitation.")

Execution Variables

These are our suggestions of alternate ways of doing tasks, given different circumstances. For example: "If the program is expecting a very high volume of cases, a computer consultant should be hired to program a tracking system."

Risk Control Variables

These are suggestions of alternate methods of performing tasks which you could call "extra insurance." They are designed to help guarantee performance, such as: higher quality, more on-time completions, or more on-budget results. You should weigh these alternatives carefully, because they often mean increasing workloads for someone in the program or outside of it, and they may lead to higher program costs or to resistance by outsiders to working with the program. Example: "Using overqualified contractors may reduce risk; however, it costs additional funds."

Other Considera-tions

These are comments that "flag" issues of concern related to the immediate step or stage. An example: "The contractor's evaluation of the program is important information used to determine if the program should revise current policy or procedures."

Getting Started

To build a new system, first skim through both the main text and the addenda of the guide, noting stages, steps, tasks, and the corresponding forms.

Next, begin developing working procedures, stage by stage. Don't jump around -- the stages occur chronologically. You'll find that the First Stage, "Program Planning," outlines steps for designing your program from scratch. We have found that the experiences of real projects are the best teachers for refining tasks, procedures and documentation, and that is why we suggest you use real cases in the planning process.

If you already have your program up and running, you may want to use the manual to "troubleshoot" and redesign areas of your program that are causing problems, usually cost overruns or log jams.

Single-Family Selective Rehabilitation is not intended as a doctrine that should be applied in all places at all times. Rather, it is designed to be a model -- a template, really -- for low-income housing programs to help them develop their own process manuals and thereby increase production.

EXAMPLE 1: IDEALIZED STAFFING PATTERN FOR SINGLE-FAMILY SELECTIVE REHAB

Number Units/Year	10-50	50-100	100-200	200-500
Average Cost/Unit	$10,000	$10,000	$10,000	$10,000
Annual Volume	100K-500K	500K-1M	1M-2M	2M-5M

Number of Full Time Positions:

Executive Director	1/6	1/4	1/4	1/4
Program Manager	1/4	1/2	1	1
Rehab Specialist	1/4	1 - 2	5	5 to 8
Loan Officer	1/4	1 - 2	3	3 to 5
Marketing/Intake	1/4	1	2	3 to 5
Office Staff	1/4	1	2	3 to 4
Specmaster Clerk	1/4	1/2	1	1 1/2
Accounting Staff	1/4	1	2	2
Program Planner	1/4	1/4	1/4	1/4

Stage: Program Planning

The program planning stage is the most critical in this single-family pipeline model. It's the time when you design and lay out the "factory." Although it seems like a time to theorize, it's not! Only experienced Rehab Specialists, Program Planners, Program Managers, and/or Executive Directors are capable of designing the system so that it will work.

The RWG has for years designed from scratch and retuned production systems. Our judgments about procedures that are typically most effective are collected here. Adherence to these guidelines will result in a high-production system. Deviation, without good cause and complete understanding, is likely to result in increased overhead, low production, and reduced product quality.

The program set-up stage is extensive for Single-Family Selective Rehab. Once done, it's there -- house after house, year after year. It is time well spent because the cost of this set-up work can be recaptured in a single year of efficient operation.

Step 1: Evaluate Neighborhoods

The purpose of this step is to obtain enough data to make educated decisions regarding the neighborhood or neighborhoods in which you may undertake housing production. For one neighborhood or a set of neighborhoods, you will determine:

- **The critical needs for housing production.**
- **The barriers to housing production; for example, crime, environmental problems or social apathy.**

It is assumed, even in a proposed area-wide program, that all planning steps will be performed at the level of individual neighborhoods.

Work Tasks

1.1. Mark up a census tract map of an area or city with two kinds of census data:

- Number of households in poverty, and
- Number of units without complete plumbing.

Program Planner

1.2. Color code the map in both categories for:

- **Percent of Poverty** - less than 10 percent poverty households, 10 to 30 percent, and more than 30 percent.

- **Percent of Substandard Units** - less than 5 percent, 5 to 20 percent, and more than 20 percent. This can be for one neighborhood or a bigger area. These or similar measures will give you a rough idea of where the most distressed areas are.

Program Planner

1.3. **Superimpose the natural boundaries** of the distressed neighborhood(s) on the map. Usually, city planning agencies have defined such boundaries, although they should be verified with neighborhood leaders, if and when you formalize them.

Program Planner

1.4. **Draw neighborhood boundaries** for a minimum population of 4,000 to 5,000. Smaller areas are hard to organize for housing production. A serious housing production organization will want to work in a neighborhood or sets of neighborhoods with at least 250 units needing help. Otherwise, production levels will likely be inefficiently low.

Program Planner

1.5. **Do a windshield survey of "reject" factors for a neighborhood.** The following factors may lead you to reject a neighborhood for housing production at the present time. (See Addenda 1a and 1b, "Sample Housing Survey Forms.")

- Large numbers of barren lots, "snaggletooth" blocks, and vacant buildings, with no major redevelopment being planned.
- Large incursions of commercial, industrial, and/or transportation operations into a formerly residential neighborhood, so that a critical mass of housing no longer remains.
- Open drug dealing and drug use, dirty streets, poor police and fire protection, unboarded vacant buildings and other signs of crime, and little or no signs of reinvestment. [Note: These will be fruitful neighborhoods for reinvestment only after community organizing has occurred and critical services are on the upswing.]

Program Planner

1.6. **Determine if the neighborhood has significant public or "special" resources.** Sometimes a recent economic or political change, such as a new factory, the abandonment of a road plan, or a public redevelopment plan, will spell an inevitable improvement in a neighborhood, despite other negative factors. Note these in your analysis, but consider them significant only when there are real commitments of resources (talk is cheap). If "improvement" only means gentrification so that middle-income families are drawn into the neighborhood, the end result will be the displacement of the neediest families.

Program Planner

1.7. **Decide what income groups you want to work for.** Income-group priorities are among the most important planning decisions a group can make, and these decisions can have a large impact on production models you choose. This manual generally assumes that the greatest need is with very low-income households, and that some income mix is beneficial. We suggest you prioritize the neediest households and set quotas for the following classifications of families by income, since these are U.S. census categories.

- Families below 50 percent of median income
- Families at 50 to 80 percent of median income
- Families over 80 percent of median income.

Note: If quotas are not set, programs tend to avoid the neediest households. Definitions of these ranges are available from your local Community Development office or from HUD.

Board of Directors

1.8. **Decide if the neighborhood is even feasible to work in** based on this analysis.

Board of Directors

1.9. **Make a rough allocation of resources** among neighborhoods, if this is to be an area-wide program. Prepare to be flexible if one area outperforms another.

Board of Directors

Execution Variables

- If a group has organized around a specific project for which there are identified resources, then Step 1 and several of the steps that follow will not be necessary. If this is the case it should be recognized that the group is organizing for a project and not production. Example: rehab of one historic building for which special funds are available.

Step 2: Determine Appropriate Production Model(s)

This step represents further feasibility analysis on selecting appropriate production models. It avoids unnecessary surveying and data collection for production models that clearly are not feasible. This step consists of a more detailed analysis of housing conditions. It is accomplished by driving through every street in the neighborhood, using basic census data, and doing interviews.

Work Tasks

2.1. **Determine the general character** of the housing stock, specifically:

- Total number of housing units (census data, from Step 1)
- Total units rented and owner-occupied (census data)
- Family size and composition
- Rough number of abandoned units (windshield survey)
- Rough number of obviously dilapidated units (windshield survey)
- Rough number of vacant, for-rent and for-sale units (count for-rent signs, read newspaper ads, talk to representatives of real estate companies, interview tenants). With rentals, try to get a sense of the shortages in one, two, three, and three-plus bedroom units. Often there are shortages of larger units but not of smaller ones, which signal the need for a focused gut rehab or new construction strategy.
- Rough rental costs by number of bedrooms, size of unit, and utility costs. High costs (over $300 to 400 a month) can indicate shortages and a need for new units -- again, sometimes of very specific sizes.
- Ballpark estimates for acquisition and construction costs for occupied, vacant and new units.

2.2. **Consider new construction or substantial rehab** of vacant structures as a strategy if:

- There is available land or buildings.
- Rental vacancies are less than 2 to 4 percent and rents are more than $300 to $400 a month for the "low-end" units. These would indicate a scarcity driving up prices.
- There is no widespread need for selective rehab as a first strategy to stop the decline of the existing stock.
- There are signs of selective rehabs and other reinvestments that have paved the way for higher levels of investment. If this is not present, new renters and/or owners may not move to "new" housing in a bad neighborhood, nor will the housing make the neighborhood "good" unless a critical mass can be built quickly.
- The cost of acquiring and rehabbing occupied housing exceeds the cost of gut rehab and new construction.
- There are vacant buildings and lots that need filling in, in a basically sound neighborhood.

Program Planner

2.3. **Develop a plan for substantial rehab or replacement of occupied units** only if there is extreme structural damage or very substandard original construction. Otherwise, selective rehab will be more cost effective and less disruptive to occupants.

Program Planner

2.4. **Develop a plan for selective rehab as a first strategy if:**

- Regardless of any other factors, including low vacancies, the housing stock is declining and being abandoned faster than it is being repaired. In this case, larger per-unit investments are not warranted until this condition is turned around.

- There are few signs of reinvestment in a neighborhood worthy of investments.
- Board-ups of vacants, crime control, and community organizing are either unnecessary or have started already.

Program Planner

2.5. **Decide on appropriate production model(s)**, on a preliminary basis, pending more detailed planning.

Executive Director

2.6. **Develop a plan for boarding up or tearing down vacant structures** in cooperation with the local government if:

- Vacancy rates are acceptable, i.e., there is no housing shortage.
- The primary need is a long-term program of "catch up" maintenance in occupied housing. Generally, the subsidy required for one new or substantially rehabbed unit will be sufficient for the repair of three or four occupied units.
- Historically or culturally important buildings need to be preserved. They can be secured for future restoration after the more immediate and widespread problems of residents have been addressed.

Program Planner

Step 3: Determine Potential Producers

This step will help you determine whether a neighborhood-based group or an area-wide group should undertake the work needed.

Work Tasks

3.1. **Estimate the volumes of units possible and needed** in each production model. It is very expensive to design and staff a multiyear program to rehab or build a few housing units. If the total production of a brand-new program with new staff is less than the following rough guidelines, your organization should ideally look to a higher-volume agency or for-profit business to carry out the work. Otherwise, the first 50 or 100 units will suffer from the inefficiencies and mistakes of your learning curve.

- New Construction/Gut Rehab - 100 units total program volume.
- Multifamily Selective Rehab - 200 units total program volume.
- Single-Family Selective Rehab - 200 units total program volume.

Program Planner

3.2. **Try to find a partner to do lower-volume production models.** This agency or business should have at least two years' successful experience with the model(s) undertaken.

Executive Director

3.3. **Failing that, gear up for a one-time project.** Try to do it quickly -- in two years or so. Line up the best people with at least three years' successful experience with the model undertaken. Don't plan to hire unskilled people.

Executive Director

Other Considerations

- **Don't commit to anybody** until you have surveyed the neighborhood and available resources. (See Steps 4 and 5.)

Step 4: Collect Data on Resources

The purpose here is to find out what resources are likely to be available to the program.

Work Tasks

4.1. **Make a list of all actual and potential loan and grant sources** for each production model.

Program Planner

4.2. **Review past public and private investment** practices in the neighborhood, or similar ones, to gauge their receptiveness.

Program Planner

4.3. **Estimate the volume** of funds that might be available for your program from these sources. "Overall shortfall" requires knowing need, i.e., pro forma.

Program Planner

4.4. **Factor in the kinds of self-help work** that already exist in the neighborhood.

Program Planner

4.5. **Factor in housing code enforcement** efforts required to sustain the improvements, particularly in rental housing. Ask the housing code enforcement agency to consider providing the number of inspectors required to perform systematic inspections of all housing on a three-year cycle.

Program Planner

4.6. **Evaluate Financing Requirements.** Investigate each possible funding source for requirements that affect your production process. Key points to evaluate include:

- The requirement for an appraisal

- Loan-to-value ratios
- Minimum housing standards
- Eligible applicants
- Eligible areas
- Eligible improvements
- Ineligible improvements
- Inspection requirements
- The total amount of money available to a single dwelling
- Compatibility with other money sources
- Homebuyer underwriting criteria
- EEO contracting rules
- Fair wage laws

Program Planner

4.7. **Get political support** for your likely production model(s). Make sure neighborhood groups and leaders, local churches, and politicians either like you, or are willing to tolerate what you are doing. Any of them can put you out of business. The main approach is explaining over and over all the reasons for your concept and incorporating feedback.

Executive Director

4.8. **Get formal or informal commitments** from the key individuals and organizations, such as local government or financial institutions, whose continual support you will need.

Executive Director

Execution Variables

- Sometimes a new public financing program comes along (e.g., Section 8 new construction, or urban development action grants in their day) that is relatively easy to access in its first year or two. This can be a springboard to get your first financing and a steady stream of funds in years to come.

Step 5: Collect Detailed Neighborhood Data

To plan your project well and to attract funding to it, you must do a careful and systematic preselection of neighborhoods and available housing stock. If neighborhoods are in the process of severe decay and housing is too deteriorated to be economically salvaged, then your efforts may be wasted. In addition, it is important to know the readiness of residents to participate in the project. It's naive to assume that everyone will eagerly accept your plans, even in the face of severe need. Some of the tasks below are optional, depending upon which production models look feasible and how much good data can be obtained from other sources.

Work Tasks

5.1. **Collect all preexisting surveys** less than 10 years old. These include:

- U.S. census (census tract data)
- Surveys by local government, neighborhood groups, service organizations or others. City housing inspection departments or community development agencies are the most likely source.

Program Planner

5.2. **Get a list of structures abandoned and with code violations,** if available. Cities with good housing inspection departments will have this.

Program Planner

5.3. **Look at previous public housing investments** in the neighborhood. Community development agencies, the HUD area office, and the local housing authority should, upon written request, give you a list of all past, present, and pending subsidized projects in the neighborhood by address.

Go look at a sample of these projects. Their condition can indicate what production models have failed or succeeded.

Program Planner

5.4. **Ask the police department** where they make frequent calls and arrests in the neighborhood, and what, if any, are the dangerous blocks to stay away from.

Program Planner

5.5. **Map out vacant lots and/or potential re-use buildings,** along with buildings that appear to need demolition. Use a "topographical" map from the City planning department, as it will generally show addresses, building outlines, and lot lines. This task is only needed if you are pursuing the gut rehab or new construction models.

Program Planner

5.6. **Map out deteriorated existing housing** on a similar topographic map, based on a windshield survey. Have at least three categories:

[1] In need of cosmetic treatment (usually paint)
[2] Badly deteriorated but salvageable
[3] Unsuitable to rehab due to obvious structural problems or exposure
to elements a long time.

Program Planner

5.7. **Do a random sample survey** of housing in the neighborhood.

- Using your topo map, define the geographic boundaries of the neighborhood you are going to survey. Within these boundaries, determine or estimate the total number of housing units and plan to survey from 2 to 5 percent of the houses, depending on your time and resources. Select a ratio, such as every 40th residential address in the neighborhood. Make sure your sample is no less than 25 units.

- Survey each house, whether occupied or not.

- Collect the following key data (see Example 2, "Sample Housing Survey Form," and Addendum 1):

    - Number of occupants

- Number of rooms and bedrooms
- Household income
- Rent or mortgage payment
- Existence of a rent subsidy
- Utilities, taxes, insurance, etc.
- Cost to rehab (by your standard)
- Rehab desired (by occupants' standards)
- Desire for a different house inside or outside neighborhood
- Willingness to pay more for rehabbed house or replacement house
- Other demographic information, as needed.

These surveys should be assigned to a consultant or staff person experienced enough to estimate cost to rehab.

Senior Rehab Specialist

5.8. **Compile the information** from the survey and other sources. Create critical reports to provide information on such areas as:

- **Overcrowding** - units with more than 1.5 people per room, or other standard. This indicates need for new construction/substantial rehab, or new housing opportunities outside the neighborhood.
- **Cost-to-rehab** - adjust your standard to include some of the "wish list" repairs; that is, repairs that your project is not prepared to undertake unless there are extra funds. Develop both an average cost-to-rehab and number of units in ranges such as $1,000 to $5,000 (minor-selective), $5,000 to $15,000 (moderate-selective), $20,000 plus (substantial). This indicates the rehab model(s) you should choose and volumes needed.

- **Cost vs. income** - Cross-tabulating income versus cost-to-rehab gives you a general picture of affordability or need for subsidy. Guidelines:

- Incomes under $7,000 indicate no additional rent or mortgage payment is generally possible.
- Incomes of $7,000 to $12,000 may handle some minor rehabs with low subsidy, but will need ever-deeper subsidies for more costly work.
- Incomes of $12,000 to $20,000 should indicate an ability to use more market-rate money for selective rehab and more moderate subsidies for gut rehab and new construction.

- **Ability to pay** - High or low existing housing costs, as well as willingness to pay more, will cause one to adjust the guidelines above. Whether housing costs are now "high" is very situational. Rural people who have built their own house might consider 20 percent of income high for housing costs. People accustomed to paying 40 percent of income for rent might still pay a little more. And in general, the poorer the people are, the less percentage of income they can devote to housing.
- **Mismatches** - Scan actual individual surveys and reports to find patterns. For example, look for very low-income renters in substandard but historic houses that would be highly expensive to rehab and require great subsidy, while older residents are selling decent houses at a much lower cost. A strategy decision here could be to rehab and sell the "easy" houses and landbank the historic buildings until the market has turned around. Your strategies should encourage opportunities for residents to move to appropriately sized and priced housing. They should not necessarily try to remedy all current problems within a neighborhood by construction. Another way of putting it is: Custom-tailoring solutions for individual households is much more expensive than simply putting appropriate products into the market, and giving needy people better opportunities.
- **Estimate of participation factor** - The most common mistake of housing program planning is the assumption that 100 percent of needy families will participate in your program. Need is often instantly equated with program goals. This isn't practical. People's willingness, their desire for new or rehabbed housing, and the friction between what they want versus what your program wants will result in a "participation rate" for

whatever production model you invent of something like 20 percent to 80 percent of the need. Generally, the 80 percent is only obtained quickly by granting money for all the improvements. A target of 50 percent -- with some financial or other participation by needy households -- is a more reasonable estimate, and will encourage self-reliance and spread scarce subsidies to more families. In time, peer pressure may lead to higher participation. (See Example 3, "Typical Survey Report," and Addendum 2.)

- **Market rents and housing prices** - The survey, augmented by a trip to the tax assessor's office, will give you a sense of the range and maximum rents and sale prices. These maximums should be considered your ceiling for feasibility analysis, unless you have "write-down" subsidies for rents or sale prices.

Program Planner

EXAMPLE 2: SAMPLE HOUSING SURVEY FORM

1st try day/time__________________Go back time/day____________

HOUSEHOLD SURVEY

Interviewer(s) RMS Date Complete: 1/12/90

Dwelling units: 1 Address: 1111 Main St.

FIRST, I WOULD LIKE TO ASK SOME QUESTIONS ABOUT THE HOME YOU LIVE IN. YOUR ANSWERS WILL BE CONFIDENTIAL AND USED TO PLAN HOW BEST TO HELP YOUR NEIGHBORHOOD WITH HOUSING.

A. Do you own, rent, or have some other arrangement?

1 Own
(2) Rent or lease
3 Other (Specify): ____________
99 NR/DK (No Response/ Don't Know)

2
(1)

B. How much do you spend each month for housing?

RENTER

Rent per month

1	Under $50	(5)	$201 to $250
2	$51 to $100	6	$251 to $300
3	$101 to $150	7	$301 to $400
4	$151 to $200	8	$401 and up

- OR -

HOMEOWNER

House Payment (including taxes, or add in an estimate of taxes)

9	Under $50	14	$251 to $300
10	$51 to $100	15	$301 to $400
11	$101 to $150	16	$401 to $500
12	$151 to $200	17	$501 and up
13	$201 to $250		

- OR -

Renter/Owner

18 Nothing
99 NR (No Response)

5
(2)

EXAMPLE 3: TYPICAL SURVEY REPORT
CHART #2: OWNER COMPLAINTS BY TYPE OF HELP NEEDED
(CELL CONTENTS, NUMBER OF RESPONSES, COLUMN PERCENT)

	NO HELP	LOAN	GOOD CONTRACTOR	MORE INCOME	MATERIALS	FAMILY HELP	GRANT	OTHER	TOTALS
TOO SMALL	0	0	1	2	0	0	0	0	3
	0	0	50	18.18	0	0	0	0	
NEEDS PAINT	2	0	0	1	1	0	1	0	5
	18.18	0	0	9.09	25	0	12.5	0	
NEEDS	1	0	0	0	0	0	0	0	0
MINOR REPAIR	9.09	0	0	0	0	0	0	0	
PLUMBING	0	1	0	2	1	0	1	1	5
	0	11.11	0	18.19	25	0	12.5	100	
WIRING	0	2	0	1	0	0	1	0	4
	0	22.22	0	9.09	0	0	12.5	0	
KITCHEN	0	1	0	1	0	0	1	0	3
	0	11.11	0	9.09	0	0	12.5	0	
BATHROOM	0	1	0	0	1	0	0	0	2
	0	11.11	0	0	25	0	0	0	
PORCHES	3	0	1	2	1	0	0	0	7
	27.27	0	50	18.18	25	0	0	0	
ROOF	1	0	0	1	0	0	0	0	2
	9.09	0	0	9.09	0	0	0	0	
WINDOWS	0	1	0	0	0	0	1	0	2
	0	11.11	0	0	0	0	12.5	0	
HEATING	0	0	0	0	0	0	1	0	1
	0	0	0	0	0	0	12.5	0	
OTHER	4	3	0	1	0	1	2	0	11
	36.36	33.33	0	9.09	0	100	25	0	
TOTALS	11	9	2	11	4	1	8	1	47

Risk Control Variables

- If your organization and potential funders have not reached some consensus on construction standards, your survey will be somewhat ambiguous. Key areas for agreement are:
  - What kinds of buildings should be targeted for repair or demolition.
  - What is the standard for selective rehab. (Agencies that only practice substantial rehab and new construction may have an approach to even selective rehab that is two or three times more costly than the standards advocated in this manual. Conversely, they may not recognize the need for selective rehab and respond only to abandoned buildings.)
  - What is the standard for gut rehab and new construction. Square footages, amenities and design treatments are key areas that can affect your economic model.

Step 6: Adopt Appropriate Production Model(s)

Up to this step, this manual has been neutral as to which production model you might choose. In this step, we expect you will formally adopt one or more of the following production models:

1) Single-Family Selective Rehabilitation,
2) Multifamily Selective Rehabilitation, and/or
3) Substantial Rehabilitation/New Construction Model.

Work Tasks

6.1. **Re-evaluate your earlier decision** on which production model(s) make sense and approximate volumes possible both annually and ultimately.
Executive Director, Program Planner

6.2. **Write an outline/program plan** based on your information. Make some workable cost and financing scenarios that include:

- Acquisition costs (if any)
- Site development costs (if any)
- Construction costs
- Soft costs
- Named sources of financing
- Some model household budgets for different ranges of income, including monthly income, rents and utilities paid, payment of preexisting housing debt (if any), potential payment on new housing

debt, utilities, taxes, insurance, and maintenance. These will obviously vary for owners and renters. Some guideline income ranges to look at are: up to $7,000, $7,000 to $12,000, and $12,000 to $20,000. These target per-unit subsidy demands, a critical limitation of any program.

Board of Directors, Executive Director

6.3. **Formally adopt the plan.**

Board of Directors, Executive Director

6.4. **Obtain formal or informal commitments** from organizations with resources most critical for your program. Ideally you have gone to them before. Now you have a written plan that may vary from what you said previously. Incorporate as much of their feedback as you can.

Executive Director

6.5. **Think big.** If you don't have the resources you need, go ask for them. Likely candidates:

- People or organizations who have already made financing or "soft money" investments in the neighborhood and who might be impressed by a coherent plan.
- City-wide, area-wide, state-wide, or national organizations that agree essentially with your goals and have probably seen a number of unprofessional attempts to attain them. Examples (in descending order of likelihood): community development agencies, local corporations and foundations, state housing finance agencies, and national foundations and corporations with an interest in community development in your locale.

Executive Director

Step 7: Set Program Construction Standards

It may be "one for the money, two for the show" in some games, but a successful selective rehab program starts by defining the "show," or what a typical safe and decent house would look like.

Only after this standard has been negotiated and adopted by all lenders, building inspection departments, and program staff can the cost of an individual rehab be accurately estimated. (See Example 4, "Typical Program Construction Standards," and Addendum 3.) These decisions can be very difficult if large numbers of poor households are to be served with always insufficient funds.

◙ Work Tasks

7.1. **Review the model standards** included in Addendum 3. The senior staff of the program should review these model standards for modification and application to their own program and city.

Senior Rehab Specialist

7.2. **Price out options for the local market.** Using Specmaster, create a prototype rehab house called "Standards Cost." Price the total cost of bringing the prototype rehab house (unit) up to the recommended minimum standards. List other "optional" improvements that might be allowed. Using the model standards, for example, the prototype might cost $9,350; an upgraded standard allowing some occupant or board requested "wish list" items might cost $12,500 per unit. It is absolutely mandatory to use comparative price information when making standards decisions.

Senior Rehab Specialist

7.3. **Review Government and lenders' minimum standards.** Identify conflicts between the proposed program standards and Government and lenders' minimum standards. Set up meetings with local jurisdiction and lender officials to discuss a flexible approach to housing renovation. Use the price information developed for the prototype rehab house to show the impact of the upscale standards, and negotiate for minimum standards closer to the model standards.

Senior Rehab Specialist

7.4. **Prepare recommendations.** In asking for approval from the board of directors for this standard, clearly state the economic impact of each decision concerning housing standards. Use a presentation format that identifies the cost differences among several optional standards. Recommend a minimum standard to the board, and identify the cases in which that minimum standard would be exceeded.

Senior Rehab Specialist

7.5. **Adopt the standard for the SFSR program.** Finalize the standards as approved by the board of directors. Use these standards in the Single-Family Selective Rehab program as a guide for all rehab staff and as a guideline for clients.

Executive Director

Risk Control Variables

- Not creating or revising program construction standards will often result in a loss of control over the scope of work and the resulting budget, because funders, government programs, and concerned citizens will apply their individual standards. This almost always results in trying to build Cadillacs on a Volkswagen Beetle budget.

Other Considerations

- Higher standards also result in higher maintenance costs. The cost of maintaining 10 feet of kitchen cabinetry will always be higher than for four feet. The same logic applies to calculating replacement reserves.

EXAMPLE 4: TYPICAL PROGRAM CONSTRUCTION STANDARDS

Heating System Standards

Heating systems - Inoperative, hazardous or inefficient (under 60 percent efficiency per a test) heating systems will be repaired to perform at 78 percent or better efficiency or replaced. Replacement space heating systems are acceptable where they are preferred by the occupants and have a lower rehabilitation cost and the same or lower operating cost as compared to central heat. All heating equipment will have a minimum expected useful life of 10 years, if preexisting, or 20 years, if new. Fuel tanks will be located in fire-resistant areas. Habitable rooms must be provided with 68° heat when the outside temperature is at the average yearly minimum for the locale.

Wood and Coal Stoves - Wood and coal stove installations will have safe clearances from combustible surfaces as defined by the local Fire Marshal. Protective floor and wall coverings will be provided as necessary. Flues will be safe, properly sized and clean.

Electrical System Standards

Electrical outlets - Every room will have at least one duplex receptacle and one light fixture or 2 duplex recepticals. Large rooms or those with heavy current electrical loads may require three or more outlets, as determined by the program inspector. Kitchens will have at least two outlets on separate circuits, one of which must have 20 amperes capacity, and one of which may be 15 amperes. Bathrooms will have at least one grounded outlet.

Electrical wiring - "Knob-and-tube" and other older-type wiring will be left in place if safe and serviceable. Where areas are being insulated, knob-and-tube will be removed and all junction boxes covered. Recessed ceiling light fixtures will be covered before insulation is placed over them.

Ground fault interrupters - Electrical outlets within 4 feet of a bathtub must be removed, or replaced with a GFI-protected outlet. Any newly wired outlet in a bathroom or within 5 feet of a kitchen sink area must be GFI protected.

Light fixtures - Every habitable room must be provided with a light fixture or one outlet controlled by a wall switch. Wall or ceiling lights with their own internal switches are acceptable. Common or internal stairways from one floor used as a living area to another must have at least one light controlled by a 3-way switch, or in the case of public hallways, a 3-way switch or automatic switching device. Exterior stairs over five risers will have a light fixture.

Step 8: Develop Financing Forms and Guidelines

Once you know what a prototype house costs to rehab, you must integrate these costs into the proposed financial vehicle(s). Using real cases and proposed financing, we can set up a few financing prototypes for the Intake and Rehab departments to use as guidelines for eligibility, and to develop application forms.

Work Tasks

8.1. **Research** the following information from all funding sources:

- The requirement for an appraisal
- Loan-to-value ratios
- Minimum housing standards
- Eligible applicants
- Eligible areas
- Eligible improvements
- Ineligible improvements
- Inspection requirements
- The total amount of money available to a single dwelling
- Compatibility with other money sources
- Homebuyer underwriting criteria
- EEO contracting rules
- Fair wage laws

The program financing department will do most of this research. Construction considerations should not be overlooked. Once the information is collected, a typical or prototype range of cases involving different types of financing should be developed from this information.

Loan Officer

8.2. **Develop fill-in-the-blanks worksheet** relating to the most common financing methods available. (See Example 5, "HUD's Section 312 Loan Program Checklist.")

Loan Officer

8.3. **Apply the worksheets to three or four "real" cases.** Using the prepared worksheets and previously developed construction standards, inspect three or four real cases as similar as possible to the client group the program proposes to serve. The three or four cases should provide data with which to develop typical construction budgets, financing fees, and other soft costs on these jobs.

Senior Rehab Specialist

8.4. **Establish parameters of construction.** For each loan source, establish a minimum and a maximum construction cost for the Rehab Specialist as guidelines for doing preliminary inspections. This prevents the rehab specialist from specifying $35,000 worth of work when the maximum loan is $18,000, or specifying $8,000 worth of work when the lender's standard would require $9,500.

Senior Rehab Specialist, Loan Officer

8.5. **Develop clear guidelines for rejection** of applications based on feasibility, income eligibility, and appraised value.

Program Manager, Loan Officer, Senior Rehab Specialist

8.6. **Maintain current information on all loan and grant sources** and apprise the rehab specialists, in writing, of all changes affecting the lender's construction standards, underwriting criteria, and maximum loan values.

Senior Rehab Specialist, Loan Officer

8.7. **Develop an application form** (see Addendum 4) that gathers all the needed data about household income and demographics.

Intake Staff, Loan Officer

Risk Control Variables

- Financing prototypes should include an ample contingency for construction.

Other Considerations

- It is important to the construction department that other departments also use the financing prototype information. It is especially useful to the Intake Department in eliminating ineligible clients before their rehab jobs are specified and costs estimated.
- Lenders in well-established state and federal programs may be able to provide financing application forms or worksheets that can be put to use without modification.

EXAMPLE 5: LOAN PROCESSING CHECKLIST 1-4 UNITS
OWNER OCCUPIED

Name(s) ______________________________
Address ______________________________
Soc. Sec. #(s) ____________________________

1. Interest rate determinations [CH. 2-3.]
   a. Income $____/mo.
   b. Number in family
   c. Median income adjusted for family size $_______
   d. Qualifies for ____% interest rate

2. Amount for GPI $____/ loan amount $____= ____%
   [Not to exceed 40%] [Ch.3-2.]
   [GPI / cost of rehabilitation if refinancing]

3. Loan limitations [Ch.3-2.]
   a. Loan requested . . . . $_____
   b. Limitation per unit $33,500 . . . . $_____
   c. Cost of rehabilitation . . . . $_____

4. Owner occupied under 80% of median income [3% loan][Ch 7-5.c.]
   a. Property value after rehabilitation . . . . $_____
   b. Existing debt . . . . + $_____
   c. 50% to 80% of 312 loan [Ch 7-5.c (3)] . . . . + $_____
   d. Total debt . . . . = $_____
   e. Maximum 312 Loan - - 80% of 4a . . . . $_____

5. Owner occupied over 80% of income [CH.7-5. c.]
   a. Property value after rehabilitation . . . . $_____
   b. Existing debt . . . . $_____
   c. 312 loan amount . . . . + $_____
   d. Total debt . . . . + $_____
   e. Maximum 312 loan amount -- 90% of 5a . . . . $_____

6. Loan limit if not refinancing [least of 3a, 3b, 3c, 4e, 5e] . . . . $_____

7. Loan limit if refinancing [Ch. 3-3]
   a. Amount to be refinanced . . . . $_____
   b. Lesser of 3c and 4e . . . . + $_____
      [5e if borrower's income is 80% to 95% of median]
   c. Maximum loan if refinancing . . . . = $_____

8. Term of loan [Ch. 2-2.]
   a. Term requested by applicant . . . . ____ years
   b. Economic life after rehabilitation
      ______years X 75% = ... ____ years
   c. Maximum loan term [lesser of 8a and 8b] . . . . . . ______years

9. Refinancing eligibility [Ch. 3-3.]
   a. Applicant's income . . . . $_____
   b. 95% of median income . . . . $_____
      Eligible [only if 9a is less than 9b]
      Yes __ No __
   c. Monthly payment on existing loan [P & I] . . . . $_____
   d. Monthly payment 312 loan [P & I based on line 3c] . . . . + $_____
   e. Total monthly mortgage payments [P & I] . . . . + $_____
      Monthly payment ratio [line 9e ./. 9a] . . . . %_____

Step 9: Customize the Inspection Checklist

Once you have installed a well-designed software package into your computer, your staff has mastered how to run it, and the specifications that come with the package have been customized to your own projects' needs and local pricing, the software can save 50 percent or more of your time. The program will enable you to produce a variety of documents, such as: the scope of work for contractors, cost estimates, bid documents, and work write-ups. In this section, we refer to Specmaster, which is the proprietary specification and estimating software developed by The Enterprise Foundation. The same tasks apply to all specification and estimating software.

The key to simple and efficient use of Specmaster, or other specification software, is the creation of a field input form, or checklist, which rehab specialists can use to record scope of work, locations, quantities, and unusual conditions. This "tool" must undergo refinement to make it of maximum value for your use, and it must be field-tested prior to full application to the pipeline.

◙ Work Tasks

Revise the checklist that comes with Specmaster software (see Example 6, "Typical Checklist Edits," and Addendum 5) based on the program's last 10 cases or on staff knowledge of field conditions.

Senior Rehab Specialist

9.1. **Take this revised checklist into the field** and inspect three to five typical units. If your projects vary substantially from one building type to another, it may be necessary to repeat this step more than once as you change designs for each new type.

Senior Rehab Specialist

9.2. **Add missing specifications.** Handwrite a two or three word "slug," a short name for a specification, on the checklist when a required spec is not listed. This is needed because of expected variations in local standards and construction practices. Put the normal unit of measure next to each slug being used.

Senior Rehab Specialist

9.3. **At the office, look up the spec numbers** of the handwritten slugs in the software's index. If none exist, assign new spec numbers to the new slugs. Write these numbers on the checklist as they are assigned.

Senior Rehab Specialist

9.4. **Revise the checklist** based on the first round of inspections, add the new specs with new numbers to the checklist, and delete ones that are now obviously not applicable.

Senior Rehab Specialist

9.5. **Retest the Checklist.** Return to the field and use these second generation checklists to inspect an additional three units of each type. If 15 percent of the specs (slugs) must be handwritten, then repeat Work Tasks 9.2. (creating new specs and slugs) and 9.3. (revising and deleting specs after an additional round of inspections). The goal is to create a usable list that contains 90 percent of the specs used on any job in the program.

Senior Rehab Specialist

9.6. **Print out the final checklist** and use it to inspect all program jobs for the next six months.

Senior Rehab Specialist

Other Considerations

- There is an extremely strong tendency to continue to use a yellow pad instead of the input sheets when starting to use Specmaster. If the yellow pad prevails, it will add 15 to 20 percent additional administrative time to the spec-writing process.

EXAMPLE 6: TYPICAL CHECKLIST EDITS

BATHROOM ________ **by** ________ **;** ________**Floor**

PLUMBING

22- 6240	________	EA	Shut Off Valve
22- 6242	________	EA	Supply Line and Shut Off
22- 6279	________	EA	Plumbing -- Custom $
22- 6442	________	EA	Faucet Repair
22- 6460	________	EA	Lavatory-Wall Hung Complete
22- 6448	________	EA	Dual Control Faucet
22- 6450	________	EA	Secure Wall Hung Lavatory
22- 6466	________	EA	Vanity 24" Complete
22- 6488	________	EA	5' Steel Tub w/Shower Complete
8- 4130	________	EA	Tub end Wall
22- 5422	________	EA	Tub Surround – Fiberglass
22- 6470	________	EA	Diverter & Shower Ring
22- 4416	________	EA	Shower Curtain Rod
22- 6486	________	EA	Water Saver Showerhead
22- 6512	________	EA	Replace Commode
22- 6506	________	EA	Repair Commode (Guts)
22- 6712	________	EA	~~Lavatory & Commode~~ 2 piece bath - All Work
22- 6718	________	EA	~~Lav+Commode+Tub/Sh.~~ 3 piece bath – Complete - All Work
__- ____	________	__	________________

ELECTRIC

23- 7018	________	EA	GFI Receptacle
23- 7072	________	EA	Replace Switch
23- 7074	________	EA	Replace Outlet
23- 7094	________	EA	Replace Light Fixture
23- 7092	________	EA	New Light Fixture & Switch
23- 7090	________	EA	Light Fixture Globe
23- 7066	________	EA	Electric Repair -- Custom $
__- ____	________	__	________________

SHEET GOODS AND FLOOR TILE

19- 5540	________	~~EA~~ SF	Underlay and VCT
19- 5524	________	SF	Underlay & Vinyl Sheet Goods
19- 5504	________	SF	Tile flooring - Repair

WALLS AND CEILINGS

15- 5134	________	SF	Gut and Drywall - 1/2" W C
15- ~~5122~~ 5124	________	SF	~~Shellcoat~~ Laminate Drywall - ~~3/8th~~ 1/2" W C
15- 5104	________	SF	Drywall - Renail and Retape W C
15- 5004	________	SF	Patch Plaster -- Large Area
8- 4290	________	LF	Shoe Mold
8- ~~4286~~ 4283	________	LF	Baseboard - ~~1x4~~ vinyl
8- 4162	________	EA	Trim Door Set - 1x4
8- 4166	________	EA	Trim - Window Set 1x4
__- ____	________	__	________________

Step 10: Customize the Specmaster Database

Construction specifications must reflect the individual program's standards, material and method preferences, local codes, and achievable costs. Specmaster Software, which is recommended in conjunction with this Process Manual, comes preloaded with specs and costs but these must be edited and added to, to reflect local conditions.

Work Tasks

10.1. Review preloaded costs and specs. Assign your most experienced cost estimator to review the unit prices and the Specmaster database. Review the last 10 jobs under contract if the program already uses itemized bids. Review the database for missing specifications, wording that does not describe the way the program does its work, and the trade names. The trade names should follow the way work is subcontracted in your locale. This makes it easier for contractors to "sort out" the work.

Senior Rehab Specialist

10.2. Delete/add specs. Remove specifications that are never used in your locality by your contractor pool. Add any specifications that are missing, including a normal unit measure, a short name (slug), and performance specs. These tasks are best done on the first actual projects undertaken with Specmaster. The key to efficient and successful use of Specmaster is adapting the specifications and costs to match local practice.

Senior Rehab Specialist

10.3. Approve draft database. The senior rehab staff person should approve all the prices, units, and specifications to be used by the program.

Program Manager

10.4. Test the database. Inspect three to five units. Use the same inspection data that was collected for checklist creation.

Senior Rehab Specialist

10.5. Input the data from three to five inspections as a new project with a new project name.

Specmaster Clerk

10.6. Edit the work write-up with prices and create text for the new specs. Enter these changes into the database through the menu, "Specification Maintenance." (See Example 7, "Index of a Customized Specmaster Database," and Example 8, "Typical Edits to a Database.")

Senior Rehab Specialist, Specmaster Clerk

10.7. Print a combined "Cost Estimate" for each job. It will show prices for the specs used in the dry run. Examine the unit price, that is, the individual price of each item, or specification. These prices should be verified and changes written on the report.

Senior Rehab Specialist, Specmaster Clerk

10.8. Enter these local revisions into the database ("Specification Maintenance"). Retest the database - inspect 5 to 10 additional units and make any necessary changes.

Specmaster Clerk

10.9. Adopt the program database.

Senior Rehab Specialist

10.10. Print a record set of the documents.

Specmaster Clerk

10.11. Save the database on floppy disks, label them carefully and file them.

Specmaster Clerk

EXAMPLE 7: INDEX OF A CUSTOMIZED SPECMASTER DATABASE

ITEM	SPECIFICATION	TRADE	UNITS	LABOR	MATERIALS	SUB-CONTRACTORS
14	Finish Carpentry					
3500		14	EA	0.00	0.00	0.00
3501	**Windows**	14	EA	0.00	0.00	0.00
3502	Sash Lock	14	EA	0.00	0.00	12.00
3504	Window Lift	14	EA	0.00	0.00	8.00
3506	Sash Cords	14	EA	0.00	0.00	12.00
3510	Window Sash-repair	14	EA	0.00	0.00	25.00
3514	Window Repair	14	EA	0.00	0.00	25.00
3516	Window Repair-Basement	14	EA	0.00	0.00	0.00
3518	Window Weather-strip	14	EA	0.00	0.00	20.00
3522	Window Sill-Replace	14	LF	0.00	0.00	25.00
3524	Window Stool	14	EA	0.00	0.00	30.00
3528	Window Repair-Custom	14	EA	0.00	0.00	0.00
3530	Window Trim Set	14	EA	0.00	0.00	37.00
3534	Wood Window Sash-Replace	14	EA	0.00	0.00	65.00
3536	Fixed Sash-Replace	14	EA	0.00	0.00	55.00
3537	Awning Sash	14	EA	0.00	0.00	50.00
3540	Window-Basement Unit - Replace	14	EA	0.00	0.00	110.00
3544	Window-Replace Double Hung	14	EA	0.00	0.00	180.00
3560	Window Aluminum Screen & Frame	14	EA	0.00	0.00	20.70
3564	Window Screen	14	EA	0.00	0.00	4.00
3568	Window Screen-Custom	14	EA	0.00	0.00	0.00
3570	Reglaze Sash	14	EA	0.00	0.00	15.00
3574	Glass Replace-Wood Sash	14	EA	0.00	0.00	17.00
3578	Glass Replace -Aluminum Sash	14	EA	0.00	0.00	16.00
3580	Glazing-Store Front	14	SF	0.00	0.00	18.00
3582	Glass Replace-Skylight	14	SF	0.00	0.00	5.00
3586	Glass Block	14	SF	0.00	0.00	20.00
3588	Glass Replace-Custom	14	EA	0.00	0.00	0.00
3600		14		0.00	0.00	0.00
3601	**Cabinets**	14		0.00	0.00	0.00
3602	Cabinets-Repair Base & Wall	14	LF	0.00	0.00	20.00
3610	Cabinet-Metal Base With Sink	14	EA	0.00	0.00	500.00

EXAMPLE 8: TYPICAL EDITS TO A DATABASE

Spec Number :	5834	UCI Number :	.- -
Unit Labor :	0.000	Unit Material :	0.000
Unit Sub :	$2,800	Cost Per Unit :	0.000
Trade :	21	Units :	EA
Slug :	FURNACE & DUCT--65,000 BTU		
Spec :	Dispose of existing furnace to legal dump. Install a ~~Sears~~ TRANE or preapproved equal, 65,000-BTU, gas-fired, forced-air furnace with a minimum AFUE rating of ~~80~~ 82 and distribution system including metal surface mount duct work, registers, set back thermostat, power wiring and flue piping.		

Spec Number :	5840	UCI Number :	.- -
Unit Labor :	0.000	Unit Material :	0.000
Unit Sub :	~~1200.000~~ $875	Cost Per Unit :	0.000
Trade :	21	Units :	EA
Slug :	FURNACE--30,000 BTU GAS FLOOR		
Spec :	Install a gas-fired, 30,000-BTU floor furnace including all gas line, thermostat, flue piping and grills to effect a complete, code-approved installation.		

Spec Number :	5850	UCI Number :	.- -
Unit Labor :	0.000	Unit Material :	0.000
Unit Sub :	475.000	Cost Per Unit :	0.000
Trade :	21	Units :	EA
Slug :	FURNACE--35,000 BTU GAS WALL		
Spec :	Install an AGA-approved upright, surface mount, 35,000-BTU wall furnace. Include a direct-vented flue pipe, gas piping ~~and~~ thermostat, and		

Execution Variables

- If the organization does not have computers and cannot get access to computer time from other sources, a low-volume system can be created using a word processor, a set of standards, a performance manual of specs, and a word processing template. This template would include the major repetitive specs. Specs on individual projects can be made by deleting and adding to this template, and including quantities.

- If sweat equity is planned, one requirement is specialized specifications which list tools, equipment, and materials necessary to complete a specific task. These specs should be loaded into the trade called "Sweat Equity."

Other Considerations

- Adapting the database is a tedious, detail-oriented job that must be completed prior to any professional and accurate use of Specmaster.

Step 11: Develop a Contractor Qualification System

Identifying and maintaining a pool of cost-effective and capable construction workers is essential to rehabbing homes. This rehab process requires three actors: a property owner, a rehab specialist, and a general contractor. Cultivating a qualified pool of contractors will solve or create problems depending on how well the job is attended to.

◙ Work Tasks

11.1. **Review the Model Contractor Classification System** (see Example 9, "Model Contractor Classification System," and Addendum 6 for a complete document). It establishes four categories of contractors based on experience, workload capacity, licenses, and past performance. In this production model the use of smaller general contractors is recommended.

Program Manager

11.2. **Evaluate the contractor market.** It is important to identify the number and type of contractors willing and able to work in selective rehab. This is particularly important for newly established programs, and programs considering a major increase in volume. In "hot" renovation markets and seriously depressed inner-city areas, it may take considerable effort to attract sufficient contractors to meet the program's demand.

Program Manager, Senior Rehab Specialist

11.3. **Evaluate state and city licensing requirements.** Obtain copies of all state and city construction licensing requirements and review them. Incorporate these criteria.

Senior Rehab Specialist, Program Manager

11.4. Implement a contractor qualification system. Create a hierarchy of contractor qualifications (see Example 10, "Contractor Qualification Procedure").

Program Manager

EXAMPLE 9: MODEL CONTRACTOR CLASSIFICATION SYSTEM

Contractors are divided into four classes depending on size, licenses, past performance, and financial capacity.

1. <u>Class 1</u>

   Limits:
   a. Jobs up to $10,000.
   b. Contractors may not carry more than $25,000 of outstanding work.

   Requirements:
   1. Application plus successful probation period.
   2. Insurance.
   3. Continued timely completion of jobs.

2. <u>Class 2</u>

   Limits:
   a. Jobs up to $24,999.
   b. Contractors may not carry more than $45,000 of outstanding work.

   Requirements:
   1. Application plus successful probation period.
   2. Insurance.
   3. Continued timely completion of jobs.

3. <u>Class 3</u>

   Limits:
   a. Jobs to $50,000.
   b. Contractor may not carry more than $75,000 of outstanding work.

   Requirements: All Class 1 and 2 plus -
   1. State License.
   2. Successful completion of five jobs or $75,000 of outstanding work.
   3. Financial capacity to carry $12,000 per week of work.

4. <u>Class 4</u>

   Limits:
   a. Jobs - unlimited.
   b. Up to $200,000 of outstanding work.

   Requirements: All Class 1, 2, & 3 plus -
   1. Financial capacity to carry $20,000 per week of work.

EXAMPLE 10: CONTRACTOR QUALIFICATION PROCEDURE

1. Contractor completed application form and filed with program

2. Contractor furnished certificate(s) of insurance

3. Program staff reviewed documentation

4. Program or lending institution staff verified information

   a. Creditor/business information

      - Dun and Bradstreet status reviewed
      - Suppliers interviewed
      - Subcontractors reviewed and interviewed

   b. Review of contractor's previous work

      - On-site visit(s) conducted
      - Telephone interview with past customer(s) completed

   c. Better Business Bureau contacted

   d. Consumer Protection Office contacted

5. Program staff reviewed contractor's file and decision was made for inclusion or rejection of contractor on Bidders' Probationary List

6. If awarded a contract, the Senior Rehab Staff reviewed first job

7. Program staff decides to accept or reject contractor for Permanent Bid List

Execution Variables

- If sweat equity is contemplated, establish criteria for owner skills and experience required for individual specifications. Different tasks will require different experience and skill. For example, almost all owners can paint. Rebuilding masonry walls is an activity very few owners are qualified to perform.

Risk Control Variables

- Requiring performance bonding of the entire contractor pool will increase the probability of completing every job, but it will do so at a cost increase ranging from approximately 5 percent to 40 percent for each and every job. This results from fewer and larger contractors participating and building the bond cost into their bids. Contractors with fewer than 10 employees can have significantly lower overhead, but may shy away from or be unable to obtain bonding.
- Thorough and accurate research of a contractor's track record is the least expensive performance guarantee method, and it is the same process all bonding companies follow.
- Requiring liability and property damage insurance in excess of the state-required minimums or standard practice can be extremely expensive and drive away smaller contractors.

Step 12: Set Up a Contractor Recruitment and Evaluation System

The program should recruit a number of contractors, forming a pool of bidders to perform the rehab work. Maintaining that pool requires ongoing recruitment effort and developing a sufficient number of profit-producing jobs to keep contractors interested. One important part of contractor management is establishing a filing and tracking system, and establishing methods for evaluating contractor performance.

Work Tasks

12.1. **Estimate the annual number and value of general contracts** that will be required by the program.

Program Manager

12.2. **Estimate the number of jobs** in progress simultaneously, by season, in an average year. To do this, first estimate the average construction period. Then do a simple bar chart illustrating high and low points (see Example 11, "Analysis: Jobs in Progress").

Program Manager

12.3. **Create a list of contractors to be invited to do program work.** Use information from:

- The Yellow Pages
- Lumber yards
- Other community profit and nonprofit programs
- Fire insurance agents
- Agencies responsible for encouraging minority contractor participation

Mail out a promotional flyer, applications, and requirements (Addendum 9, "Requirements to Enter Contractor Registry," and Addendum 8, "Rehabilitation Contractor Registration Application") to four times the number of contractors needed in the pool. Aim at influencing these contractors to apply for program work and invite them to attend a meeting.

Senior Rehab Specialist, Program Manager

12.4. **Conduct the contractor recruitment meeting** and accept applications. At this evening meeting, take about an hour to explain both the program goals and the general conditions for working with the program. Give recruits an idea of how many jobs will be bid every week or month. Make a presentation designed to motivate people to join the program's pool of qualified contractors, and answer everyone's questions. Listen for feedback -- particularly aspects of your program that the contractors don't like. (Take notes and pay close attention.)

Program Manager, Senior Rehab Specialist, Executive Director

12.5. **Verify the application information.** Run credit checks on the contractors and verify past job performance. Seek reference checks from suppliers and subcontractors. Clarify any discrepancies that surface with the contractor. (For interview techniques and forms, see Addendum 10 a-f.)

Senior Rehab Specialist, Rehab Specialist

12.6. **Award a classification level** to each contractor using the system described in Step 12 (see Example 9, "Model Contractor Classification System") and add their names to the bidders' list. Mail out a notification of inclusion on the bidders' list to the accepted contractors.

Senior Rehab Specialist

12.7. **Establish a contractor file.** Create a file for each active contractor with a work-in-progress log attached to the inner cover. This log should chart:

- Job address
- Contract amount
- Contractual start date
- Actual start date
- Contractual completion date
- Actual completion date
- Total change orders and the rehab specialist on the job

Update and investigate contractor status prior to every bid award. Include job evaluations at the end of each job.

Senior Rehab Specialist

12.8. **Develop criteria for dropping contractors** from your list. These could include poor quality of work, late completion(s), nonpayment of subcontractors, high incidence of change orders, or poor behavior on jobs. Set up procedures for eliminating unsatisfactory contractors from future bidding.

Senior Rehab Specialist, Program Director

Risk Control Variables

- The larger the pool of qualified contractors, the less likely that the loss of a single contractor will damage the program. Larger contractor pools can also result in reduced construction costs due to increased competition. Keep in mind, however, this delicate balance: If there's too much competition for too little work, especially early in your program, most of your pool will sour on your program -- perhaps permanently.

EXAMPLE 11: ANALYSIS: JOBS IN PROGRESS

I. Assumptions:

1. Annual # Jobs: 208
2. Average # starts a week: 4 (208 divided by 52 weeks)
3. Average construction period: 4-1/2 weeks - range, 4-6 weeks
4. Average # jobs one contractor can carry simultaneously: 2
5. January and Spring are low and high points.
6. With current program cannot allow more than 5 jobs a week to start.

II. Slowtime period: Winter

January: 6 to 9 jobs in progress

Minimum # contractors: 3 to 5

III. Peaktime period: Spring

April 1 to May 30th: 12 to 18 jobs in progress

Minimum # contractors: 6 to 7

IV. Minimum Contractor Pool suggested: 10 to 14.

Step 13: Select Contract Documents

Contract documents are those that define the mutual responsibilities of the owner and contractor. The principal document is the Construction Contract. Other contract documents are referenced in the Construction Contract and are a) amendments (such as a Change Order), or b) notices or record keeping documents (such as a Notice to Proceed or Lien Waiver). These are distinguished from "management" documents used by the program for assigning and tracking work, and accounting for funds.

◙ Work Tasks

13.1. **Select the contract documents** that are required to implement the program procedures. Create new documents to meet special program needs. Modify the documents to meet lenders and/or grantors requirements. Many conditions, for example, lien laws, minimum warranty terms and minority involvement requirements, vary widely from financing source to lending source and state to state. This demands that you select from, review and customize the following types of documents:

Contract Documents	**Addendum Number**
Invitation to Bid	14
Bid Proposal Form	15
Construction Contract	17
Owner Self-Help Agreement	16
Notice to Proceed	19
Draw Request	21
Partial Release of Liens	22
Change Order	23
Final Release of Liens	26
Contractor's Warranty	28

Program Manager

13.2. Establish a Dispute Management System. Construction field problems should be solved with escalating levels of authority to avoid jumping straight to court. A typical internal process might require a Rehab Specialist/Owner/Contractor negotiation, then a Rehab Manager ruling, followed by informal arbitration by a board comprised of a contractor, a past program participant, and the Executive Director. Formal arbitration and the courts are the last level of authority. Participation in this system should be a condition of the contract.

Program Manager

13.3. Modify the documents to meet legal requirements of the state. Have all drafts of documents reviewed by the program's attorney. Local law often requires changes in the following areas: release of liens, bidding documents, construction contract, and escrow agreement. Make the modifications suggested by the attorney.

Consultant

13.4. Adopt the documents. Add to your own draft process manual sample completed documents, along with copies of the blank documents. Have the documents printed for future use. Schedule and conduct a meeting to brief the staff on policies, procedures, roles, and responsibilities.

Executive Director

Execution Variables

If you will be using the construction management mode instead of, or in addition to, general contractors, it will be necessary to modify minor language in the documents.

Risk Control Variables

- Ten percent retainage of payments due contractors is an industry standard. Increases in the retainage percentage not only eliminate smaller contractors, but have not been proven to guarantee job performance. The reduced cash flow hurts large and small contractors.

Step 14: Develop an Internal Record-Keeping and Tracking System

Sooner or later everyone wants to know "where the money went" and "who's doing what." An efficient job data recording system provides both types of information without superfluous data. It does two things: 1) records client and production data, and 2) tracks progress of jobs. The recommended system uses department checklists for details and tasks and an overall tracking system based on 7 to 12 milestones.

◙ Work Tasks

14.1. Establish client and production data categories. Have all the program departments create "wish lists" of items of information they would like to have available in various reports. Typical information categories include: sources of funding, neighborhood, census tract, client family composition, family income, political jurisdiction, whether or not the client is a past recipient of services, cost of the job, etc. Prioritize the items on the "wish lists" and decide on the categories to be used.

Program Manager

14.2. Establish interdepartmental job milestones for a Program Log. Have all departments create "wish lists" of milestones to be used for tracking each job. It is important that there not be too many or too few milestones being tracked. Seven to twelve is about right. Each department will have a detailed File Index for each job, related to its function, to monitor progress of a job from step to step. (See Example 12, "Program Log.")

Examples of key milestones are:

- Program entry date
- Application completion/referral to Production Department
- Specifications completion date
- Financing approval date
- Job bid date
- Contractor construction completion date
- Warranty inspection date.

Program Manager

EXAMPLE 12: PROGRAM LOG
ACTIVE CASES AND COMPLETIONS SUMMARY

Week Ending: 7-15-88

	Completed Jobs or Pending Verification	Applications Rec's on Waiting List	Applications Active Cases Being Proc'd	Jobs Under Contract Loans Settled Units	Jobs Under Contract Loans Settled Amount
Section 312	38	0	54	12	$182,700
CDBG	1	53	0	26	282,727
THDA	0	0	26	0	0
CHE	0	0	0	0	0
Other	0	0	0	0	0
Rental Rehab	0	0	4	72	900,000
TOTAL	39	53	84	110	$1,365,427

TOTAL JOBS UNDER CONTRACT AND COMPLETED

Week Ending: 7-15-88

	Loans Closed Out Year-to-Date Units	Loans Closed Out Year-to-Date Amount	Under Contract Completions 1988 Year-to-Date Units	Under Contract Completions 1988 Year-to-Date Amount	Completions 07/87 to 12/87 Units	Completions 07/87 to 12/87 Amount
Section 312	24	$ 416,358	36	$ 599,058	9	$120,250
CDBG	108	859,791	134	1,142,518	25	202,412
THDA	5	133,659	6	163,659	7	221,039
CHE	0	0	0	0	0	0
Other	0	0	0	0	0	0
Rental Rehab	0	0	72	900,000	0	0
TOTAL	137	$1,409,808	248	$2,805,235	41	$543,701

14.3. **Develop departmental tasks** that will become a File Index, or checklist, for each job (see Example 13, "Sample Production Department Job File Index"). Here, instead of 7 or 12 activities, there may be 50. The Production Department File Index identifies every typical step a job might require of the Rehab Specialist. The index is fastened in front of the departmental file folder. Certain milestones, like "job bid date," will be posted both in the File Index and the Program Log. Not only the Production Department, but Intake, Finance, and Accounting will have File Indexes, or checklists, to mark progress of cases. This is essential data to move a case that's bogged down in the system and to assess employee productivity. One last file needs a File Index -- the Program Record Set of Documents File, called the Record Set File.

Program Manager

14.4. **Develop a Contractor's File Index** that serves as a checklist for application and qualification documents. It also lists the jobs a particular contractor is awarded, their start and completion dates, and copies of evaluations.

Project Manager

14.5. **Review and customize the system.** Finalize the Log and Indexes. Put all current jobs on this system and test it for three months. After the trial time, make necessary modifications.

Project Manager

Execution Variables

- Paper systems are generally recommended for job progress tracking and computer systems for client and production data. However, at high volumes (more than 150 units a year) it may be necessary to perform both functions on computers.

Other Considerations

- An effective system will need modifications as program volume changes.
- A tracking system costs money and time to develop. However, one must be in place to effectively monitor each job, monitor program and

employee performance, and provide documentation in the event of litigation or program audit.

- The Program Log and the Department File Index are usually accepted as evidence in courtrooms. Both could be particularly important to prove **if** something was done and **when** it was done.

EXAMPLE 13: SAMPLE PRODUCTION DEPARTMENT JOB FILE INDEX

BORROWER(S): ____________

ADDRESS: 1111 E. Main St.

TELEPHONE: HOME: 920-6777 WORK: (Borrower) N/A

(Co-Borrower) 920-1111

ACTIVITY DESCRIPTION		DATE REQUESTED	DATE RECEIVED	COMMENTS
Pre-Application				
Verification of Employment	(a)	6/11/89	6/12/89	
Verification of Employment	(b)	6/11/89	7/12/89	
Verification of Mortgage		6/11/89	7/16/89	
Verification of Loans(s)		6/11/89	7/15/89	
Verification of Income		6/11/89	7/12/89	
Credit Report		6/11/89		
Title Opinion				
Work Write-Up			8/25/89	
Appraisal				N. Required
Owner's Approval of Work Write-Up			8/30/89	
Bid Proposal				
Bid Review			8/30/89	
Contractor Selection by Borrower				
Completed Loan Application				
Approval by Field Office				
Loan Set-Up Form				
Transmittal to Lender (Initial Settlement)				
Executed Rehabilitation Contract				
Proceed Order				
Request for Progress				
Payment Drawdown	1)			
	2)			
	3)			
	4)			
Request for Final Drawdown				
Final Waiver of Liens				
Certificate of Final Inspection				
Closeout Documentation forwarded to Lender				

FINAL COMMENTS: ____________

Step 15: Develop Formal Program Procedures

A well-working Single-Family Selective Rehabilitation program is like an efficient manufacturing plant. Every operator knows his or her job and how it affects the total product. Establishing an overall process early on helps guarantee program efficiency.

Work Tasks

15.1. **Review this guide** in its entirety, noting the execution variables, the risk control variables and other considerations for each step in the process. Add in the marketing, intake, and financing steps in their entirety. Review the model documents. If necessary, cut and paste until you have a complete process manual with which to process single-family dwellings through rehab. Make sure the "job titles" match actual positions in your shop. Have each member of the staff, including those involved with intake, financing, construction, and policy, review the proposed process manual.

Senior Rehab Specialist, Program Manager

15.2. **Board/staff approval.** Present the process manual to your board of directors after modification by the staff to meet local requirements, and request approval.

Executive Director

15.3. **Legal review.** Submit all of the proposed documents and procedures to the organization's attorney, after the manual has been approved by the board, to review for conformance to local and city regulations and state law. Minor modifications to the documents and procedures may be required.

Consultant

15.4. Assign responsibilities. Finalize all changes in the procedure. Identify each step in the process and assign a staff person to execute that step. If the job titles in this manual aren't appropriate, they should be changed. Each staff person should be responsible for a cluster of activities within the overall process.

Program Manager

Other Considerations

- Breaking down the process into too many steps may add unnecessary costs to the program.

The final form of the process manual will vary based on four main factors: volume, type of client served, extent of construction, and local trade practices.

Stage: Design and Feasibility

Hand writing and reproducing specifications is a practice of the 1950's. In the 1990's, jobs are inspected with a computer input sheet. Specifications and cost estimates are then generated by the computer.

Highly skilled technicians can now outperform three of their 1950's colleagues. Two hours after the field inspection, the completed job design can be tested against financing guidelines. The Rehab Specialist position still requires knowledge of construction materials, methods, and costs, but tedious administrative tasks are minimized and the decision time is accelerated.

In this Process Manual, design terminology and design documents relate to Specmaster, a proprietary software developed by The Enterprise Foundation. There are several similar software products on the market that use different forms and terminology, but also produce specifications and estimates.

There are hundreds of estimating software products that do not produce detailed bid specifications. None of these is appropriate for the production model presented here, because of the lack of spec-writing capabilities.

Step 16: Perform Specification Inspection

The specification inspection is a detailed assessment of the building's condition. The task is to measure, photograph, and evaluate work required to rehab the building, and to enter precise data onto a Specmaster Checklist.

Work Tasks

16.1. Log the job in the File Index when assigned. Review the loan officer's range of prequalifications for each applicable loan or grant program for this job.

Rehab Specialist

16.2. Make an appointment with the client to perform a property inspection. This requires approximately two hours for the average job. All of the owners should be present to head off misunderstandings that may require a redrafting of the specifications.

Rehab Specialist

16.3. Select the appropriate checklist for the type of work involved. Bring a clipboard, several checklists, a flashlight, a measuring tape, a camera, graph paper, a set of housing standards, and a copy of the loan/grant guidelines with you.

Rehab Specialist

16.4. Interview the occupant. To break the ice, first interview the occupant or owner in the kitchen or living room. Verify all the client information on the referral sheets. If the occupant is not the owner, get the telephone numbers where each of the owners can be reached at work. Ask the occupants to describe their history with the structure and their assessment of it. Ask them what kind of improvements they want or think are required.

Rehab Specialist

16.5. **Walk through the structure and record data.** Using the Checklist, take the occupants with you and go room by room through the house, pointing out deficiencies and making entries on the Checklist for the work required to bring the structure up to standards. (See Example 14, "Filled-Out Specmaster Inspection Checklist," and Addendum 5 for an example of a Specmaster Inspection Checklist.)

Rehab Specialist

16.6. **Measure each room and take photographs of any deficiencies** when there are questions about the most effective repair to make. Note extra costs for items having unusual dimensions. These can be expressed either as a "whole dollar" increase (e.g., $20) or an increased cost per unit (e.g., $20 a square foot). These are called "price boots."

Rehab Specialist

16.7. **Add unique specifications** for items not on the checklist. A slug is a short name for a specification, such as: "3/4-INCH CDX PLYWOOD," without any descriptive language.

Rehab Specialist

16.8. **Review the checklist prior to leaving.** Sit down with the occupant before leaving and review the work marked on the Specmaster Checklist for thoroughness, completeness, and understanding. Clarify any "wish list" items and communicate which ones may or may not be financeable. If any items have been overlooked, inspect them.

Rehab Specialist

16.9. **Establish expectations** with occupant. It is important the rehab specialist and the occupant become a team as early as possible. The debriefing portion of the interview is the time to: 1) establish realistic expectations, 2) review the roles and responsibilities of each of the team members, and 3) emphasize the importance of the help the occupant can offer to the success of the rehab effort.

Rehab Specialist

Execution Variables

- In a few programs, the rehab specialist prepares a list of items requiring work or "deficiencies" and the contractor prepares the specifications in the form of a proposal to the homeowner. It is subsequently reviewed and approved by the rehab specialist. This system is generally not recommended, except where contractors can demonstrate that they can write detailed and careful specifications that conform to program guidelines.

EXAMPLE 14: FILLED-OUT SPECMASTER INSPECTION CHECKLIST

INTERIOR ROOM 12' by 10'

LR:______ DR:______ HALL:______ BR: **Front** OTHER:______

ELECTRIC

23- 7006	____	EA	Electric Cover Plate
23- 7004	____	EA	Replace Wall Receptacle
23- 7010	1	EA	Receptacle - 15 AMP Wire At front wall
23- 7072	____	EA	Replace Switch
23- 7094	____	EA	Replace Light Fixture
23- 7076	____	EA	Switch Light
23- 7080	____	EA	Switch Wall Receptacle
23- 7092	1	EA	Light Fixture and Switch
23- 7090	____	EA	Light Fixture Globe
23- 7108	____	EA	3-way Switch & Fixture
23- 7126	____	EA	Smoke Detector - Hard Wired
23- 7066	____	EA	Electric Repair -- Custom $

HVAC

21- 6173	____	EA	Register -- Wall
21- 6172	____	EA	Floor Register
21- 6102	____	EA	Heat Duct & Register
21- 6138	____	EA	Custom HVAC Repair $______
21- 6144	____	EA	Thermostat -- Mechanical Setback
__- ____	____		____________

WINDOWS

8- 3502	2	EA	Sash Lock
8- 3506	____	EA	Sash Cords
8- 3504	2	EA	Sash Lift
8- 3514	____	EA	Repair Window
8- 3574	1	EA	Wood Sash - Replace Glass left window
8- 3534	____	EA	Window Sash - Replace
8- 3544	____	EA	Replace Double Hung Window
8- 3536	____	EA	Fixed Sash - Replace
10- 2776	2	EA	Storm Sash

DOORS

8- 3410	2	EA	Rework Interior Door Closet + bedrm.
8- 4376	____	EA	Privacy Lock

Step 17: Prepare Work Write-up (With Prices)

The work write-up and cost estimate are required to determine the scope of work, document the quantity, establish the quality performance standard, and identify the construction method and the projected cost estimate of every proposed work item.

◙ Work Tasks

17.1. Review the checklist. To the extent possible, the checklist should be completed in the field. Now, prior to inputting the data, double check your data. Double check raw field measurements (e.g., repainting a 12 ft. x 14 ft. living room) and square footage entered on the checklist. (It should be 172 square feet.) Make addendum statements legible or more complete. Review the standard unit costs and modify them one last time for special job conditions that may have been overlooked in the inspection.

Rehab Specialist

17.2. Look up the numbers for slugs you write in. When the checklist doesn't have a spec you need, the index of the Specmaster database may show the slug you wrote in. Read the spec carefully to verify it is appropriate. Write the number of the spec on the checklist with the appropriate quantity and addendum.

Rehab Specialist

17.3. Custom write specs that are missing from both the checklist and index. This requires writing descriptive language for your "slug." Request a spec number from the senior rehab specialist.

Rehab Specialist

17.4. Input the checklist data. In order to reduce the margin of error, assign a specific staff member whose sole responsibility is to input information into the computer. Provide this person with a corrected and legible checklist for input.

Specmaster Clerk

17.5. Review printouts. Print out the Work Write-Up with prices. (See Example 15, "Sample Work Write-Up Report.") Review it for the accuracy of the estimate and as a check for work items that may have been forgotten. Print out a separate Cost Estimate (see Example 16, "Specmaster Cost Estimate Report"), which combines identical trade items at different locations. Review this for pricing. Mark up the hard copy of the Work Write-Up with changes. Re-input all changes and print the final copy. (See Addendum 13 for a complete example of sample Specmaster reports.)

Rehab Specialist

17.6. Have a higher level of review for difficult conditions or where the Rehab Specialist's skills are in question (e.g., a trainee).

Senior Rehab Specialist

17.7. Notify the Loan Officer of the results. Pass the Work Write-Up (with prices) to the staff person responsible for the funding decision.

Rehab Specialist

Execution Variables

- Programs without specification software can achieve a moderate level of production efficiency by using a checklist, a word-processed typical house template, and a performance manual. The checklist still acts as a field input device, but word-processed templates must be developed for each predominate type of house stock (e.g., 1930 brick duplex, 1920 three-story Victorian). The template is then modified to reflect this job's actual conditions.
- In smaller shops, the Rehab Specialist may have to input the specs and print reports. This is not recommended in high-volume shops, as a clean hand-off of the checklist to another person usually results in

more careful inspections and less design "at the computer" where field conditions are only a memory. Even if "what if" scenarios are wanted, they are most efficiently produced on paper, then handed off for input.

Risk Control Variables

- After preparation of final specifications, a 10 percent contingency is still recommended. If the project is too tight to allow for this contingency in the financing proposal, the scope of work should be modified until a 10 percent contingency is possible.

EXAMPLE 15: SAMPLE WORK WRITE-UP REPORT

Work Write-Up

Date: 08-25-89 11:02 AM

Project: CDBG Address: 1111 East Main St. Unit: 1

Item No.	Item	Quan.	Unit	Cost
	Stairs to 2nd			
18-5404	LAMINATE WITH 1/4" PANELING Cover Wall with 1/4" prefinished paneling, glued and finish nailed 8" on center. Remove loose plaster and top piece of 3-piece base. Slip plywood 1/2" under base molding and break all joints on studs. Reinstall ogee molding and 2" wainscot on finished wall. BOTH SIDES	12	SF	$123
23-7084	3-WAY SWITCH (PAIR) Install two 3-way, ivory switches at opposite sides of room at strike side of door controlling an existing fixture. Fish wire or use surface mounted raceway as directed by owner.	1	EA	$115
23-7126	SMOKE DETECTOR Install a UL-approved, ceiling-mounted smoke and heat detector permanently wired into a receptacle box outside sleeping rooms.	1	EA	$ 66
	Front Bedroom			
14-3410	DOOR--REWORK INTERIOR Rehang and plane door, and adjust door and lockset to operate properly.	2	EA	$ 84
14-3470	DOOR STOP Install baseboard mounted solid metal door stop.	2	EA	$ 10
14-3502	SASH LOCK Install brass-plated sash lock on double hung window.	2	EA	$ 10
14-3506	SASH CHAINS Install sash chains between sash and counterweights. Adjust window to operate smoothly.	2	PR	$ 34
14-3534	WINDOW SASH Install single-glazed wood sash on existing frame. Replace parting bead and stop where necessary to allow lower sash to operate. Fix top sash shut.			

EXAMPLE 16: SPECMASTER COST ESTIMATE REPORT

Cost Estimate

Date: 08-25-89 11:05 AM

Project: CDBG Address: 1111 East Main St. Unit: 1

Item No.	Item	Quan	Unit	Labor	Mater	Subs	Total
1 General Requirements							
1-4	PERMITS AND FEES	1	EA	0	0	36	36
1-54	JOB SITE SIGN	1	EA	0	0	20	20
1-70	DEBRIS REMOVAL	3	EA	0	0	300	300
	Subtotals			0	0	356	356
4 Site Work							
4-310	FENCE--CHAIN LINK 4' HIGH	33	LF	0	0	198	198
4-360	GATE--GALVANIZED 4' HIGH	1	EA	0	0	65	65
4-540	T-POLE	2	EA	0	0	8	80
	Subtotals			0	0	343	343
6 Concrete and Paving							
6-876	CONCRETE SLAB--4"	20	SF	0	0	70	70
6-882	SIDEWALK REPAIR	36	SF	0	0	126	126
	Subtotals			0	0	196	196
8 Rough Carpentry							
8-1556	HANDRAIL BRACE	2	EA	0	0	30	30
	Subtotals			0	0	30	30
9 Roofing							
9-2304	ROOF--DOUBLE COVERAGE SIS	7.5	SQ	0	0	600	600
	Subtotals			0	0	600	600
10 Conservation							
10-2776	STORM WINDOW/ SCREEN-CAULKED	9	EA	0	0	540	540
	Subtotals			0	0	540	540
14 Finish Carpentry							
14-3340	DOOR-EXTERIOR FLUSH-REAR	1	EA	0	0	150	150
14-3342	DOOR-EXTERIOR PANELED-FRONT	1	EA	0	0	250	250
14-3410	DOOR-REWORK INTERIOR	7	EA	0	0	294	294
14-3470	DOOR STOP	5	EA	0	0	25	25
14-3502	SASH LOCK	4	EA	0	0	20	20

Step 18: Perform a Sources-and-Uses Analysis

A "sources-and-uses" analysis, a comparison of available funds with estimated costs, is performed to ensure that a job either can be completed within budget or should be abandoned as unfeasible.

◙ Work Tasks

18.1. Compare actual funding available (sources) to the cost estimate (uses). The funding available should be the amount the program's Finance Specialist or Loan Officer has established as maximum loan or financing available for this client. The cost estimate should include construction, plus fees, construction financing charges, and all other "soft" costs.

Rehab Specialist

18.2. Revise the scope of work of the job if the cost estimate is over budget. (This is called "rescoping.") You may propose lower-priced materials or a less costly construction method, or delete certain work. Make multiple passes through this rescoping task to attempt to bring a job within budget. If you succeed and the work is substantially modified, the owner's approval of these changes should be obtained, preferably by telephone. If a work write-up cannot be developed within budget, it should then be referred to program management to make the decision to seek additional funding or abandon the job as unfeasible.

Rehab Specialist

18.3. Log the completion of the Work Write-Up in the Program Log and Job File Index.

Rehab Specialist

◙ Execution Variables

- Adding sweat equity can reduce money requirement.

Step 19: Secure Preliminary Project Approval

Preliminary project approval is the important step of seeking and gaining consensus among all parties to proceed with a financing commitment and bidding.

◙ Work Tasks

19.1. **Mail the final Work Write-Up with prices** to the owner.

Rehab Specialist

19.2. **Gain final owner acceptance of the Work Write-Up.** Unless there have been substantial changes since the site inspection, this can be done over the telephone.

Rehab Specialist

19.3. **Review case.** Conduct a full review with the Finance Specialist or Loan Officer and the Senior Rehab Specialist now that the job is in "final" form and ready to progress to the bid stage.

Program Manager

19.4. **Note any needed changes or refinements** in program policy or procedures posed by the job, and submit them for review and implementation as soon as possible. If there are minor and acceptable variations, approve the job "as is" and introduce the changes in policy or procedures starting after this job. If major modifications to policy or procedures are required, the job must recycle back to the spec-writing step.

Rehab Specialist

19.5. **Give the owner preliminary financing approval** by mail, subject to bidding.

Program Manager

19.6. Approve the job for bidding.

Senior Rehab Specialist

19.7. Post preliminary approval to the Program Log and Job File Index.

Rehab Specialist

Stage: Bidding

Bidding and getting contracts is still a very personal process. Even after you've standardized an exceptional set of construction management documents, vouchers, schedules, and contracts, the best guarantee of performance is the commitment of well-chosen contractors to their jobs.

Of course, overbearing and inconsistent demands for documentation or "jumping through hoops" will hamper the process and scare away some contractors.

The goal is to reach a balance between the qualities of a good "handshake deal" and rigorous documentation used in military procurement regulations. You are going to be responsible for balancing the interests of the property owner, the contractor and your own program. Your program will be asking for the most paperwork to be completed, so try to keep your demands to a reasonable minimum.

This stage can be simplified into three basic steps:

[1] A notice of work to be done
[2] Offers by contractors to do the work
[3] An agreement between the owner and one of the contractors.

In this time of frequent law suits, well-written instructions for an established construction program are still the best defense against misunderstanding that can lead to litigation.

Step 20: Match Potential Bidders to the Job

This step builds upon Step 11: Develop a Contractor Qualification System. This is the practical application of procedures intended to match qualified contractors to each job. At the very least, these procedures will determine the minimum acceptable contractor for a job.

Work Tasks

20.1. Review the Program Log. Identify the contractors involved in other rehab jobs who may be qualified and capable of accepting additional work. Review the bidders' list to identify other eligible contractors to bid on a job.

Rehab Specialist

20.2. Review the Contractor Files. Make note of performance problems. Exclude from immediate consideration of additional work any contractor with whom the program is experiencing current problems.

Rehab Specialist

20.3. Identify any "special requirements" of the job. What contractor qualification level does this job require? Does this job require a contractor with specialized skills? Does this job require a contractor with the ability to get higher-than-normal performance guarantees?

Rehab Specialist

20.4. Match bidders to job. Select at least two, but no more than six, qualified and acceptable contractors to receive Invitations to Bid. The higher number is only acceptable when the specifications are so clear to contractors they can bid "from their desk" or after a cursory inspection.

Rehab Specialist

Execution Variables

- Contracts can be negotiated. This may be appropriate on an undesirable job where time is of the essence, on a very small job, or as a general policy to save contractors and your program the overhead of bidding. Negotiated contracts are also a good way to utilize new or less-skilled contractors. With negotiated contracts, you will definitely want to agree on line-item "unit" pricing. In this variation you limit contractor qualification to identifying the contractor and initiating a request for proposal. With this strategy you must have excellent cost estimating and attempt to tell the contractor what prices you want. You must have a good negotiator. If you are routinely using negotiated contracts, bid at least every fifth job to keep everyone honest. **Note:** If a dispute should arise, you will not have any support from other bidders who may have kept copies of the specifications that were sent out to bid.

Risk Control Variables

- Using overqualified contractors may reduce risk; however, it costs additional funds.
- Bidding jobs to too many contractors with too much red tape can reduce interest and/or actually raise bid prices. A contractor's time is money, too.
- If negotiated contracts are used, be extremely careful about file documentation.

Step 21: Prepare Bid Documents

Preparing the bid documents requires filling out and/or modifying the documents selected in Step 14 in the Program Planning Stage. This ensures the formal bid package is complete.

Work Tasks

21.1. Fill in the Construction Contract document to the extent possible (see "Agreement Between Owner and Contractor," Addendum 17). The contract should clearly spell out responsibilities for loss of materials or tools from the site and for repairing vandalism.

Rehab Specialist

21.2. Fill out an Invitation to Bid (see Example 17 and Addendum 14, "Invitation to Bid"). This document should include:

- Inspection dates
- Start and completion dates
- All required performance guarantees
- Any specialized requirements of this job.

Generally, lump-sum bids which consolidate all costs should be requested, rather than bids by line item, in order to keep "red tape" to a minimum.

Rehab Specialist

21.3. Mail the bid packages to selected contractors with a courtesy copy to the owners. Record the bid list in the Job File and in each contractor's file. The package includes:

- Invitation to Bid
- Bid Proposal Form (see Example 18, "Bid Proposal Form," and Addendum 15)
- Forms required by funders (federal wage regulations, minority participation rules, lead paint requirements, etc.)
- Detailed Specs by Trade (See Example 19, "Sample Bid Page")
- Any drawings or additional specifications
- A sample construction contract form adopting all of the above items by reference.

Rehab Specialist

21.4. Enter the dates that invitations to bid were mailed in the Program Log and Job File Index.

Rehab Specialist

21.5. Answer all inquiries about the scope of work. The Rehab Specialist must be available to clarify the scope of work. He needs to respond to questions concerning acceptable material substitutions, acceptable methods of completing the work, and potential change order items. If this leads to changes in the specifications, the job should be rebid or an addendum sent to all bidders.

Rehab Specialist

Execution Variables

- Sweat equity and "Owner GC" arrangements are *not* recommended unless special procedures (and usually special staffing) are in place. If the owner is acting in place of a general contractor, then a special Owner Self-Help Performance Agreement (see Addendum 16 for a complete example of this document) should be filled out. Note: This opens the door to a **very different production model** which is not treated in this Manual.

- When starting a program, and annually thereafter, it is advisable to have several dozen jobs bid on a line-item basis. And that way, a wealth of data can be gathered in order to analyze and revise Specmaster prices.

- In well-oiled programs with repeat contractors, you may shorten the Invitations to Bid by **referencing** long boilerplate documents rather than including them in entirety, such as the contract and financing requirements. The minimum that must be included are:

  - Invitation to Bid
  - Bid Proposal Form
  - Work Write-Up
  - Drawings and Additional Specifications.

EXAMPLE 17: INVITATION TO BID

DATE: 8/25/89

JOB ADDRESS: 1111 E. Main St.

PERSON TO CONTACT TO GAIN ENTRY:

NAME: Mrs. Homeowner
ORGANIZATION: N/A
PHONE: 920-1111
BIDS DUE: 9/3/89

SUBMIT BIDS TO: Rehab Specialist

Dear Contractor,

You are invited to submit a proposal for rehabilitation as specified in the attached Work Write-Up Report. The property must be carefully examined and all quantities and field conditions verified. Please submit the enclosed "Bid Proposal Form" before 12 noon on the due date showing the following:

- Bid price for entire job.
- Earliest starting date.
- Proposed completion date.
- Any addenda showing work not covered on our Work-Write Up Report.

The program reserves the right to reject any oral bids and require any bidder to unit price any accepted proposal.

EXAMPLE 18: BID PROPOSAL FORM

TO: ______________________ DATE SUBMITTED: 8/30/89

PROGRAM MANAGER,
PROPERTY ADDRESS: 111 East Main St.
Arlington, VA

The following proposal is hereby submitted covering all Work shown and/or specified in the attached bid documents received for this job, which are:

1. WORK Write-Up Dated 8/25/89
2. ______________ Dated ______
3. ______________ Dated ______
4. ______________ Dated ______

The undersigned, in having become thoroughly familiar with the terms and conditions of the proposed Contract Documents and with local conditions affecting the execution of the Work, proposes to complete the Work for the sum of Ten thousand dollars $ 10,000 .

Within the following time frame:

Start within 10 days of notice to proceed and complete within 45 calendar days thereafter.

Upon Request by the AGENCY, contractor agrees to itemize any/or all aspects of this proposal.

Contractor: Urban Renovation R.M. Santucci (Signature)

Address: 127 S. Highland Street
Arlington, VA 22204 8/30/89 (Date)

Phone: 703-979-3707

Employer ID #: 54-0070-631

Proposed Addenda:

The Contractor further proposed the following changes, deletions, or additions to the Write-up.

	PRICE
______________	$______
______________	$______
______________	$______

EXAMPLE 19: SAMPLE BID PAGE

Detailed Specifications by Trade

Date: 08-25-89 10:56 AM
Project: CDBG Address: 1111 East Main St. Unit: 1

Item No.	Item	Quan	Units	Unit Price	Total Price
16 Drywall					
5104	DRYWALL--RENAIL & RETAPE Renail or screw gypsum surface 6" on center. Scrape seams and nail pops. Retape where tape is missing, damaged or bubbled. Finish with 2 coats of compound. Dining Room	130	SF	____	70
5124	LAMINATE 1/2" DRYWALL Install 1/2" gypsum drywall over existing plaster surface with screws 8" on center and a bead of construction adhesive 20" on center. Butt drywall to door and window casing. On walls remove top molding from 3-piece base and reinstall after surface is paint-ready. Tape and 3 coat finish. AT WALL BETWEEN MIDDLE AND FRONT BEDROOMS AND BATH DOOR WALL AND CEILING	380	SF	____	600
5162	DRYWALL--FUR, HANG, FINISH Fur surface with 1"x 3" wood or metal furring strips. Hang, tape and 3 coat finish 1/2" drywall, glued and nailed or screwed 8" on center. KITCHEN	160	SF	____	250

Bidder U. R. Co. Inc. Total $920

Step 22: Conduct Bid Review and Clarification

This step involves analyzing the bids for completeness, responsiveness, and price.

Work Tasks

22.1. **Record all the bids** received in the Production Department Job File Index and in the Contractor's File Index (see Step 14 and Addendum 11).
Rehab Specialist

22.2. **Compare bids** to each other noting dollar differences and the contents of any addenda that accompany the bids. Compare the two lowest bids to the Work Write-Up (with prices).
Rehab Specialist

22.3. **Request clarification** from the contractor on unexplained high bids or proposed variations in the start time and complete time.
Rehab Specialist

22.4. **Review each contractor's file** for work-in-progress, insurance, and timely completion of current work load.
Rehab Specialist

Execution Variables

- If this is a negotiated contract, compare the overall budget and the unit prices to the in-house cost estimate. Discrepancies should be highlighted for negotiation with the contractor.

Risk Control Variables

- Any contractor submitting a bid 20 percent or more below the final estimate, who cannot clarify the reason(s) for the deviation, should be eliminated from the bidding. If your cost estimating is accurate, that contractor won't even cover his or her own overhead expense and is headed for cash flow problems.

Step 23: Select and Notify Contractor

Select the contractor with whom the owner will negotiate the final details of a construction contract based on the completeness, responsiveness, and price of the bid submitted, and notify the contractors of the owner's choice.

◙ Work Tasks

23.1. Prepare a written Bids Comparison including a recommendation for bid selection. Mail to the owner (see Example 20, "Bid Comparison Results," and Addendum 18).

Rehab Specialist

23.2. Solicit and receive owner approval of the bid and contractor if the project is clearly within budget.

Rehab Specialist

23.3. Notify all bidders of the acceptance of the bid by the owner (include a copy of the bid comparison) and inform the owner and contractor of the next steps in the process.

Rehab Specialist

EXAMPLE 20: BID COMPARISON RESULTS

TO: Homeowner DATE: 8/30/89

FROM: Rehab Specialist CLIENT #: ______

RE: 1111 East Main Street ARL. VA

PROGRAM ESTIMATE: $10,275

The following bids were received at Program Office and opened on 8/26/89, at 4:00 PM.

NAME OF BIDDER:	AMOUNT OF BID:
ABC Contracting	$11,510
Urban Renovation Inc.	$10,000
Downhome Repair Company	$12,000

Persons Reviewing Bids:

Joe Rehab Specialist

We have reviewed each bid listed above, and we recommend Urban Renovation Inc. to perform the work on 1111 East Main Street property.

______ (Homeowner's Signature) ______ (Co-Homeowner's Signature)

______ (Date) ______ (Date)

If there are any questions about the recommendation process, please call Joe Rehab Specialist of Rehab Program, at (111) 113 - 1234.

Step 24: Perform Final Sources-and-Uses Analysis

The final sources-and-uses analysis determines whether or not projects that have been bid over-budget can be brought within the owner's budget or the limits of available financing by: a) utilizing an alternative course of action during construction, or b) securing additional financing. Otherwise, the project will have to be abandoned.

Work Tasks

24.1. **Compare the accepted bid to the project budget,** operating within the established policy that every budget should contain a nonnegotiable 10 percent contingency. If the project is only slightly over budget, i.e., 5 percent contingency remains, rescoping can be deferred to the pre-construction walk-through, discussed in the next section.

Rehab Specialist

24.2. **Identify all cost reduction options** where the accepted bid is more than 5 percent over-budget and review these options with the owner and program. These options include: reducing the scope of work, using less costly materials, utilizing a less costly construction method(s), or employing sweat equity.

Rehab Specialist

24.3. **Negotiate with the contractor and owner** to reduce the bid by discussing cost reduction alternatives, and then getting them to accept a new Work Write-Up.

Rehab Specialist

24.4. Receive the revised bid from the contractor based on the new Work Write-Up and cycle through the sources-and-uses analysis until the bid is accepted or the decision is made to abandon the job.

Rehab Specialist

24.5. Make sure revised specifications are clearly referenced in the bid. Having several sets of specifications "floating around" is an invitation to disputes or lawsuits. Date all new Work Write-ups and use that date and the number of pages as a reference in bid documents or the contract.

Rehab Specialist

Execution Variables

- Request cost reduction alternatives as bids from contractor.
- Instead of making a new Work Write-Up, a short addenda can be used to delete, change, and/or add specifications. This is only recommended where the changes are clear and simple.

Step 25: Conduct Pre-Construction Walk-Through

The pre-construction walk-through is conducted to verify that the contractor, the owner, and the rehab specialist have the same understanding of the scope of work. Wherever there is disagreement, negotiation occurs. This may involve changing the bid price, changing the Work Write-Up, or both. If agreement cannot be reached, rebid the job or cancel it. The pre-construction walk-through also provides an opportunity to review the roles and responsibilities of each of the actors during construction, identifying problems that may arise and possible solutions.

◙ Work Tasks

25.1. Set up an appointment with all owners, the contractor, and all the subcontractors, if used.

Rehab Specialist

25.2. Walk through the house with the Work Write-Up, calling on the general contractor and appropriate subcontractors, if available, to define their work to the owner. Negotiate differences and reach agreement. If changes or additional items are required to reach agreement and maintain a 10 percent contingency, revise the Work Write-Up and/or the bid price. If the contractor is reluctant to proceed given the job conditions, rebid the job. If the owner cannot be satisfied given the contractor's description of the work to be done, either rebid or cancel the job.

Rehab Specialist

25.3. Make final revisions to the standard contract, if any, reviewing these items in particular:

- Start and completion dates.
- Total price. (Don't allow increases that would "break the budget" without approval of the Program Manager.)
- Insurance and site security responsibilities.
- Allowance items (e.g., $12 s.y. of sheet goods; $55 l.f. of cabinets).
- Owner selections (e.g., sky blue trim, white base; silver gray carpet by Monsanto).

Rehab Specialist

25.4. Review the owner's and contractors' roles and responsibilities during the construction process including responsibilities for site security, particularly for losses of tools and materials or vandalism. This is especially important with vacant buildings.

Rehab Specialist

25.5. Review the permits that will be required to perform the work.

Rehab Specialist

25.6. Require the contractor to post all construction permits on the job prior to work start. This includes trade permits for electrical, plumbing, grading, etc., for which there may be no legal posting requirement.

Rehab Specialist

Execution Variables

- The pre-construction walk-through may not be needed at this stage with negotiated contracts, if the tasks were done earlier.
- If the project is minimally over-budget, the pre-construction walk-through is an excellent time to agree on cost-saving changes to the Work Write-Up, while all parties are actually looking at the work to be done. Record the agreed-upon changes and have both the contractor and the homeowner initial the handwritten revisions to the Work Write-Up.

Step 26: Have Bidder Execute Contract

This step formalizes any changes in the job since the bid was accepted. It also involves negotiating with the contractor:

[1] A formal Work Write-Up (with prices)
[2] A draw schedule
[3] A schedule of completion

In a best-case scenario, this can occur at the pre-construction walk-through; however, if the Work Write-Up has changed, an office or telephone conference will be required with the successful bidder.

Work Tasks

26.1. Factor the Specmaster Work Write-Up (With Prices) to match the contractor's bottom line. This creates a line-item cost estimate that represents the programs idea of what each item is worth. This is used as the Trade Payment Breakdown for all payments. It will be in "unit price" form, so that the cost of change orders can easily be calculated.

Specmaster Clerk

26.2. Review the prices and Payment Schedule (see Example 21, "Sample Contract Payment Schedule" and Addendum 17, "Agreement Between Owner and Contractor," for the complete document) with the contractor. The contractor must agree to, and sign off on, the proposed payment schedule and the Work Write-Up (with prices). The prices and payment schedule should then be incorporated into the contract documents.

Rehab Specialist

26.3. **Review the proposed Construction Schedule** (see Example 22, "Sample Schedule Language," and Addendum 17 for the full document) with the contractor, identifying milestones and lead times for ordering materials. Collaboratively, the rehab specialist and the contractor should establish dates for achieving each milestone and the job completion date. Enter the start date, close-in date and completion date in the contract.

Rehab Specialist

26.4. **Have the contractor execute the construction contract,** subject to owner acceptance, by a certain date.

Rehab Specialist

Execution Variables

- The contractor puts line-item prices on his/her bid proposal and submits it to the rehab specialist for approval. The danger in this is potential "front loading," or putting high values on items done earlier in the job. For unscrupulous or undisciplined contractors, this is a disincentive to complete the job.
- Using a program-mandated draw schedule, the program could, as a policy, define fixed draw amounts and fixed draw periods for jobs of certain sizes. For example, a $10,000 job would always involve three payments, at 25 percent, 75 percent and 100 percent completion, with draws not occurring more frequently than weekly.
- To help in managing large jobs or multiple small jobs by the same contractor, prepare a bar chart of the work, which will enable you to reach an agreement on job priorities and a schedule of work completion.

Other Considerations

- In discussing with the contractor the entire construction activity, plans should be reviewed pertaining to how loss of utilities, e.g., water and electric, will be handled since this can have a heavy impact on occupant satisfaction with job progress.

EXAMPLE 21: SAMPLE CONTRACT PAYMENT SCHEDULE

CONTRACT SUM/PROGRESS PAYMENTS

3.1 The Owner shall pay the Contractor for performance of the work, subject to additions and deductions by approved Change Orders, the Contract Sum of Ten Thousand Dollars. The Contract Sum is determined as follows:

Base Bid	$10,000
Addenda	-0-
Contract Sum	$10,000

3.2 Based upon invoices submitted to the Owner's Representatives, the Owner shall make payments on account of the Contract Sum to the Contractor as follows:

Draw 1	25%	$2,500
Draw 2	25%	$2,500
Draw 3	25%	$2,500
Draw 4	25%	$2,500
Draw 5		

3.3 Payments may be withheld on account of:

1. Defective work not remedied,
2. claims filed,
3. failure of the Contractor to make payments properly to Subcontractors or for labor, materials, or equipment,
4. damage to the Owner or another contractor, or
5. persistent failure to carry out the Work in accordance with the Contract Documents.

3.4 Final payment shall not be due until the Contractor has delivered to the Owner a complete release of all liens arising out of this Contract or receipts in full covering all labor, materials and equipment for which a lien could be filed, or a bond satisfactory to the Owner indemnifying him against any lien. If any lien remains unsatisfied after all payments are made, the Contractor shall refund to the Owner all moneys the Owner may be compelled to pay in discharging such lien, including all costs and reasonable attorneys' fees. Owner may withhold a retainage of 20 percent of all invoiced charges if the Contractor fails to complete all contract items upon submission of final invoice.

EXAMPLE 22: SAMPLE SCHEDULE LANGUAGE

TIME OF COMMENCEMENT & SUBSTANTIAL COMPLETION:

2.1 The Work shall commence within 7 calendar days of authorization by written Notice to Proceed from Company.

2.2 The Work shall be substantially completed no later than 60 calendar days from the date of the Notice to Proceed. The Contractor shall be liable for and shall pay the owner one hundred dollars $ 100 as liquidated damages for each calendar day of delay until the work is substantially completed.

OPTIONAL

2.3 If Work is delayed at any time by causes beyond the contractor's control, then the Contract may be extended for such reasonable time as the Owner's Authorized Representatives may determine.

Step 27: Have Owner Execute Contract and Financing Agreement

This Step is the settlement of the loan and marks a major milestone in the processing of a case.

◙ Work Tasks

27.1. **Schedule a meeting to execute the contract,** attended by the owner, Loan Officer, and the Rehab Specialist.

Loan Officer, Rehab Specialist

27.2. **Explain the financial responsibilities** to the owner, including the terms of loans and grants.

Loan Officer

27.3. **All parties should countersign the contract and financial agreement(s)** to record the construction and financial relationships of the program, owner, and contractor.

Rehab Specialist

27.4. **Log the job.** Place the signed contract in the record set folder and a copy in the Contractor's File. Notify the Loan Department of the status of the contract and have the financial agreement signed. Fill in the Program Log with contractor's name and proposed start date.

Rehab Specialist

27.5. **Set up a Work-in-Progress file as a manila folder in the Job File.** Make copies of the contract, specs, and plans. Insert blank Draw Request Forms and Change Order Forms (see Addenda 21 and 23). These copies can be written on, left in the rain, and fed to vicious dogs without the record, which is now permanently filed, being destroyed. Continue to use the regular File Index in the record file.

Rehab Specialist

27.6. **Fill out a Job Accounting Sheet.** List the contract amount, available contingency and any pre-established draw amounts using paper with columns (see Example 23, "Sample Job Accounting Sheet").

Rehab Specialist

27.7. **Get the owner's signature on the Notice to Proceed.**

Rehab Specialist

27.8. **Give the owner copies** of all documents.

Rehab Specialist

EXAMPLE 23: SAMPLE JOB ACCOUNTING SHEET

Job Accounting Sheet
Address 1111 East Main St.
Contractor Urban Renovation Consultants

Date	Payments	Remaining to Complete	Change Orders	Contingency Balance	Notes
10/1/89		$10,000		$1,000	Job start
10/12/89	$2,500	$ 7,500	-----	$1,000	
10/24/89	$2,500	$ 5,000	#1 - $100		
			#2 - $250		
			#3 - $400	$ +250	
11/6/89			#4 - $750 (pending)	$ -500	Negotiating for a deduct change order
11/10/89			#5 + $850	$ -500	
11/22/89	$5,000	$0	#6 - $250	$ +100	
11/28/89				$ +100	Transferred to Financial Dept for Prepayment

Stage: Construction

Up to this point it has all been thinking, talking, writing, planning, negotiating, and contracting for the work which is our goal. If we have done our design job well, we are now a secondary actor as the contractor begins fulfilling the owner's needs and requests.

Our job is three-fold:

[1] To help maintain job momentum
[2] To process necessary paperwork
[3] To negotiate design and funding problems discovered during construction.

Sloppy earlier work and poorly designed programs will yield an unwelcome harvest of re-doing work steps, dissatisfied owners, and over-budget, never-ending jobs.

Step 28: Issue Notice to Proceed

The Notice to Proceed gives official notice from the program and owner to the contractor to proceed with the construction on a specific date.

◙ Work Tasks

28.1. Mail two copies of the Notice to Proceed to the contractor with a program self-addressed stamped envelope. Instruct the contractor to sign and return one copy to the program in the envelope provided. Put one in his files (see Example 24, "Notice to Proceed," and Addendum 19 for prototype documents).

Rehab Specialist

28.2. Post in the File Index that the Notice to Proceed was mailed.

Rehab Specialist

EXAMPLE 24: NOTICE TO PROCEED

CONTRACTOR: Urban Renovation Inc. DATE: 10/1/89

ADDRESS: 1217 S. Highland St, Arlington, VA

PROPERTY ADDRESS: 1111 E. Main St.

Arlington, VA

ESCROW #: 12-CDB6

Dear Mr. Santucci,

Pursuant to the Homeowner/Contractor Agreement signed on 9/3/89 __________, you are hereby notified to proceed with the rehabilitation work on the property captioned above. Upon receipt of this notice, you are responsible for the rehabilitation work on said building under the terms and conditions of the Agreement, Specifications and Plans, if any. Please be prepared to begin construction within 10 days of receipt of this notice.

Please acknowledge acceptance of this notice in the space provided below.

Mrs. Homeowner
Signature of Owner

920-1111
Owner's Phone Number

Received the foregoing Notice to Proceed on __________
Date

____
Contractor

____
Signature

____
Contractor's Phone No.

Instructions to Contractor

Please sign both copies of this notice and return one copy to:

____
Rehab Specialist

____

____
Phone No.

Step 29: Verify Job Start

Verify that the contractor is adhering to the schedule and starting the job on the date specified in the contract.

Work Tasks

9.1. **Phone the occupant** on the start date and verify that the construction permits are posted and work is under way.

Rehab Specialist

9.2. **Log the job start** in the Program Log, the Job File Index, and the Contractor's File.

Rehab Specialist

Execution Variables

- If the owner doesn't have a phone, drive to the work site and verify that the construction permit is posted and work is under way.

Step 30: Conduct Progress Inspections and Make Payments

Progress inspections are conducted to document: work in progress, the percentage of the job complete and the contractor's adherence to the project plans, specifications, schedule and other requirements.

Work Tasks

30.1. **Establish periodic draw payment inspection dates** or milestone inspection points using the draw procedure specified in the contract. (See Addendum 20, "Explanation of Draw Procedures," for detailed procedures developed by one program.)

Rehab Specialist

30.2. **Schedule appointments** for conducting the inspection with the owner and the contractor. Remind the contractor of the requirement to submit a completed Draw Request form (see Example 25, "Draw Request," and Addendum 21) to the program at least 24 hours prior to an inspection date.

Rehab Specialist

30.3. **Perform a work-items-completed inspection** using general specs by location (with prices) -- the last version with cost adjustments matching the contractor's bid. (See Example 26, "Inspection Report.") Inspect the job location by location. Note with a check mark on the specs by location, all work items that are 100 percent complete. Note all items out of spec. Note all items significantly out of phase with the Construction Schedule, and discuss them with the contractor.

Rehab Specialist

30.4. **Perform field calculations** by totaling all of the completed work items on the specifications sheet. Review all items not in conformance with the specifications and prepare recommendations to the owner. Compare the contractor's Draw Request to the calculation of work completed to date.

Rehab Specialist

30.5. **Communicate recommendations** to owner and contractor. Walk through the property with both parties, explaining where the work has progressed and where improvements in construction may be required.

Rehab Specialist

30.6. **Sign the Draw Request** and establish the next draw milestone or date.

Rehab Specialist

30.7. **Log the approved draw on the Job Accounting Sheet** and request payment from the Accounting Department.

Rehab Specialist, Loan Officer

Risk Control Variables

- If you want the ultimate in risk control, you may want to require a Partial Release of Liens from the contractor prior to each draw (see Addendum 22, "Partial Release of Liens"). This is recommended only on larger jobs (more than $20,000) and/or with untried contractors, or with a marginal contractor pool. It is expensive for the contractors to do the runaround work. If you require it, make sure it's in the contract.

EXAMPLE 25: DRAW REQUEST

Borrower: Mrs. Homeowner
Address: 1111 E. Main St.
Contractor: Urban Renovation

CONTRACTOR REQUEST FOR PARTIAL PAYMENT Draw # 1 $ 2500

I hereby certify that I have completed the work specified below in accordance with the Rehabilitation Contract and request partial payment of $ 2500 per the attached itemized request. I hereby certify that all payments due my subcontractors, suppliers, and laborers have been made, and subcontractor and material house release of liens are attached.

R M Sanfucci — Authorized Agent for Contractor

10/12/89 — Date

BORROWER APPROVAL

I hereby accept all work completed by the contractor listed above and authorize partial payment of the amount specified in accordance with the above-referenced Rehabilitation Contract, as due and payable to the contractor, from the proceeds of my loan.

Mrs. Homeowner — Borrower's Signature

10/12/89 — Date

REHAB SPECIALIST APPROVAL

Total Amount Approved: __________
Retained: 10%
Net Amount Approved: __________
Progress Payment: __________

Inspector: __________
Title: __________
Date: __________

EXAMPLE 26: INSPECTION REPORT

General Specifications by Location

Date: 08-25-89 11:04 AM
Project: CDBG Address: 1111 East Main St. Unit: 1

Item No.	Item	Quan	Unit	Cost
General Requirements				
1-4	PERMITS AND FEES	1	EA	$ 36
1-54	JOB SITE SIGN	1	EA	20
1-70	DEBRIS REMOVAL	2	EA	200
25-7626	LEAD PAINT HEPA VACUUM	1	DU	200
Front Exterior				
6-882	SIDEWALK REPAIR	36	SF	126
10-2776	STORM WINDOW/SCREEN--CAULKED	4	EA	240
14-3342	DOOR--EXTERIOR PANELED-FRONT	1	EA	250
14-4340	HOUSE NUMBERS	1	EA	26
24-7504	RAILING--WROUGHT IRON--STEPS	1	EA	150
1st Floor Hallway				
14-3410	DOOR--REWORK INTERIOR INNER FRONT DOOR	1	EA	42
15-5004	PATCH PLASTER	40	SF	60
17-5254	DECORATE VACANT ROOM	1	RM	95
19-5518	UNDERLAY AND VCT IN VESTIBULE NEW SUBFLOOR	8	SF	16
Living Room				
14-3410	DOOR--REWORK INTERIOR	1	EA	42
14-3470	DOOR STOP	1	EA	5
15-5004	PATCH PLASTER	6	SF	35
17-5254	DECORATE VACANT ROOM	1	RM	95
23-7072	SWITCH REPLACE IN HALL.	2	EA	24
Dining Room				
14-3506	SASH CHAINS	1	PR	17
14-3574	GLASS REPLACE--WOOD SASH	1	EA	17
16-5104	DRYWALL--RENAIL & RETAPE CEILING	130	SF	97
17-5254	DECORATE VACANT ROOM	1	RM	95

Step 31: Perform Change Order and Contingency Inspections

Change Order and Contingency Inspections verify that it is necessary to modify the initial contract in one of three areas: completion date, scope of work, or cost.

Work Tasks

31.1. **Inspect the job** upon the receipt of a Change Order request (see Example 27, "Change Order"). Define with the contractor and the owner the nature of the requested Change Order. Verify need and proposed solution. Verify quantities and methods.

Rehab Specialist

31.2. **Negotiate the cost and time increase requirements** with the contractor and owner. Determine if funds are available with which to implement the Change Order and recommend the owner's verbal approval.

Rehab Specialist

31.3. **Execute the Change Order.** Using a standard form (see Addendum 23) fill in time changes, money changes, and scope of work changes, and have both contractor and homeowner sign it. Then approve the Change Order by affixing your signature.

Rehab Specialist

31.4. **Post the Change Order** to the File Index and notify accounting of the budget revision. Debit contingency on the Job Account sheet and credit the contract account (see Example 28, "Job Accounting Sheet").

Rehab Specialist, Accounting Clerk

Other Considerations:

- Even if no money is involved in a change, a Change Order must be executed. Each change is a modification of the contract and must be documented and approved.

EXAMPLE 27: CHANGE ORDER

FILL IN

DATE: 10/24/89 CHANGE ORDER # 1, 2, 3

JOB ADDRESS: 1111 E. Main St.

ORIGINAL CONTRACT AMOUNT: $ 10,000

ADJUSTED CONTRACT AMOUNT: $ 10,750

ITEMS TO BE DELETED FROM SPECIFICATIONS

SPEC NO.	DESCRIPTION	AMT. DELETED
	TOTAL DELETIONS NA	$ 0

ITEMS TO BE ADDED TO SPECIFICATIONS

SPEC NO.	DESCRIPTION	AMT. ADDED
15-1340	Bedroom Door Replace	$ 100
6-7701	Drywall Kit Ceiling	$ 250
18-9028	Fiberglass Tub Surround	$ 400
	TOTAL ADDITIONS	$ 750
	NET CONTRACT CHANGE	$ 750

EXAMINED AND APPROVED BY:

Rehabilitation Specialist

Contractor

Homeowner

Date

EXAMPLE 28: JOB ACCOUNTING SHEET

Job Accounting Sheet
Address 1111 E. Main St.
Contractor Urban Renovation Inc.

Date	Payments	Remaining to Complete	Change Orders	Contingency Balance	Notes
10/1/89		$10,000		$1,000	Job start
10/12/89	$2,500	$7,500	______	$1,000	
10/24/89	$2,500	$5,000	#1 - $100		
			#2 - $250		
			#3 - $400	$+250	
11/6/89			#4 - $750 (pending)	$-500	Negotiating for a deduct change order
11/10/89			#5 + $850	$+350	
11/22/89	$5,000	$0	#6 - $250	$+100	
11/28/89				$+100	Transferred to Financial Dept. for Prepayment

Step 32: Conduct Substantial Completion Inspections

Substantial completion inspections verify that the construction work has been completed to a state of possible use, and approved by the local jurisdiction's building and/or mechanical inspectors.

Work Tasks

32.1. Receive notification of substantial completion from the contractor.

Rehab Specialist

32.2. Send the owner a Notice of Final inspection and schedule an appointment with the owner and the contractor (see Example 29, "Notice of Final Inspection and Final Payment Approval," and Addendum 24).

Rehab Specialist

32.3. Send the contractor a Final Draw Request (see Example 30, "Final Draw Request," and Addendum 25) detailing the releases of liens, inspection certificates, and warranties expected at the final inspection.

Rehab Specialist

32.4. Inspect the property using the Work Write-Up. Mark off all completed items and note all incomplete items. Verify that all permit and controlled work has been inspected and signed off, including plumbing, electric, and HVAC.

Rehab Specialist

32.5. Prepare a punch list of work items. A punch list is an itemized account of all incomplete or unacceptable work items. When work is completed, it discharges the contractor's or subcontractors' obligations under the contract. Final payment is contingent upon the completion of the punch list. (See Example 31, "Typical Punch List Items," and Addendum 27.)

Negotiate an agreement between the owner and the contractor for a completion date. Have both parties sign and date the punch list.

Rehab Specialist

32.6. **Provide close-out forms,** e.g., the Final Release of Liens forms (see Example 32, "Final Waiver of Lien and Contractor's Affidavit," and Addendum 26) and the Contractor's Warranty (see Example 33, "Sample Contractor's Warranty," and Addendum 28).

Rehab Specialist

EXAMPLE 29: NOTICE OF FINAL INSPECTION AND FINAL PAYMENT APPROVAL

______________________	______________________
PROPERTY OWNER	**CONTRACTOR**
______________________	______________________
ADDRESS	**ADDRESS**
______________________	______________________
CASE NO.	**PHONE NO.**

This is a notice of final inspection of the repair work performed on your property. The contractor has indicated that all the work required by the work write-up and in the contract agreement has been satisfactorily completed. If you disagree with the contractor that all the work has been satisfactorily completed, now is the time for you to advise us and the contractor of any omissions, problems, or defects in the work. If you certify to us that the work is complete and in accordance with the construction contract, with no omissions, problems, or defects, we will issue a check payable to you and the contractor from the escrow account making final payment for the work. By signing this form, you are certifying that you have inspected the property and that the contractor's work has been completed in accordance with the construction contract.

Thank you for your cooperation in the repair of your property.

Sincerely,

Rehabilitation Supervisor

I acknowledge receipt of this notice of final inspection, and I certify that all work has been completed in accordance with the contract. In the event that it becomes necessary to rework or repair any of the listed items, I will contact the Contractor.

______________________	______________________
DATE	
______________________	______________________
WITNESS	**PROPERTY OWNER**

EXAMPLE 30: FINAL DRAW REQUEST

JOB ADDRESS: 1111 East Main St.
CONTRACTOR: Urban Renovation
CONTRACTOR REQUEST FOR FINAL PAYMENT $ 5900

I hereby certify that I have completed 100 percent of the work in accordance with the Rehabilitation Contract and request final payment of $ 5,000 which includes payment for all Change Orders of $ 900 .

R.M. Sa[illegible] — Authorized Agent for Contractor

11/22/89 — Date

FINAL PAYMENT CHECKLIST

Prior to receiving your final payment you must present to The Agency Rehab Specialist the following properly completed and notarized Lien Releases, Local County or City Inspection Certificates, and warranties.

	REQUIRED	RECEIVED
Contractor Affidavit	✓	___
LIEN WAIVERS		
General Contractor	✓	___
HVAC	___	___
Electrician	✓	___
Plumbers	✓	___
Other ____________	___	___
LOCAL INSPECTION CERTIFICATES		
General Repair Permit...........	✓	___
Plumbing Final..................	✓	___
Electrical Final................	✓	___
Heating Final...................	___	___

EXAMPLE 31: TYPICAL PUNCH LIST ITEMS
(Check off missing items and note location)

ELECTRIC — Location(s) of List Items

- __ All outlets and switches work.
- __ Plates are straight and level.
- __ All outlet plates installed tight to drywall.
- __ Panel labeled for new circuits.
- ✓ All lights have bulbs. —— kitchen
- __ Final inspection completed.
- __ Batteries in smoke detectors.
- __ Other: ______
- __ Other: ______

PLUMBING

- __ All fixtures work; not chipped.
- __ Fixtures are secure to wall or floor.
- __ Air gap for dishwasher installed.
- __ Tub is caulked.
- __ Aerators/escutcheon plates installed.
- __ All gas connections working.
- __ Final permit inspection completed.
- __ Other: ______
- ✓ Other: Clean Tub

HEATING - Cooling

- __ System responds to thermostat.
- __ Warrantee and instructions provided.
- __ Duct work free of debris.

DOORS, WINDOWS

- __ All exterior doors weatherstripped.
- __ Doors checked for tightness, lasp alignment and lock function.
- ✓ Keys available. — need 2 sets to rear door.
- __ Windows have locks that tightly close.
- __ Window operates smoothly.
- __ Screen in place.
- __ Glass clean.

FINISHES

- __ Paint removed from hardware, glass.

- __ MISC.
- __ Trash removed.
- __ Other: ______

EXAMPLE 32: FINAL WAIVER OF LIEN AND CONTRACTOR'S AFFIDAVIT

STATE OF (Virginia)ss: Escrow # ______
COUNTY OF (Arlington) Payout # 5 - Final

TO WHOM IT MAY CONCERN:

WHEREAS the undersigned has been employed by Mrs. Homeowner to furnish renovation work for the premises known as 1111 E. Main St. of which Mrs. Homeowner is the owner.

The undersigned, for and in consideration of Ten thousand nine hundred ($ 10,900) Dollars, and other good and valuable considerations, the receipt whereof acknowledged, do(es) hereby waive and release any and all lien or claim of, or right to, lien, under the statutes of the State of Virginia, relating to mechanics' liens, with respect to and on said above described premises, and the improvements thereon, and on the material, fixtures, apparatus, or machinery furnished, and on the moneys, funds, or other considerations due or to become due from the owner, on account of labor, services, material, fixtures, apparatus or machinery, furnished to this date by the undersigned for the above-described premises.

Given under my/our hand(s) and my/our seal(s) this 22 day of November, 19 89.

Witness: ______ ______

Witness: ______ Urban Renovation Inc.
(TITLE)

By: R.M. Sanfren

Title: President

EXAMPLE 33: SAMPLE CONTRACTOR'S WARRANTY

The undersigned hereby guarantees all of the work required and performed by the contract, one year from the date shown on the NOTICE OF COMPLETION as dated 11/22, 1989. Contractor also attaches herewith all manufacturers' and suppliers' written guarantees and warranties covering material and equipment furnished under the contract.

IN WITNESS WHEREOF, the undersigned has signed and sealed this instrument this 22nd day of November, 1989.

Urban Renovation
Contractor Business Name

By: R.M. Sexton

____
Witness Signature

Step 33: Perform Retainage Release Inspection

The retainage release inspection is conducted to verify the completion of all contractual agreements between the owner and the contractor and approve the release of all retained funds.

Work Tasks

33.1. **Verify punch list completion with the owner** over the phone.
Rehab Specialist

33.2. **Verify that the file contains copies** of: Final Release of Liens, long-term warranties (e.g., roof, termite treatment, all installed equipment), and the contractor's guarantee for one year.
Rehab Specialist

33.3. **Schedule an appointment** at the job with both the owner and contractor.
Rehab Specialist

33.4. **Inspect the job using the punch list.** Verify that all work has been completed.
Rehab Specialist

33.5. **Sign the final draw request releasing all retained fees** to the contractor (see Addendum 25).
Rehab Specialist

33.6. **Review maintenance procedures** with the owner and the contractor.
Rehab Specialist

33.7. **Complete the Program Log** and close-out the Job File and the Contractor's File.

Rehab Specialist

Stage: Post-Construction

Even after everyone agrees that the work is done, the job is not over. All healthy systems seek and adjust to feedback from the field. Maintaining satisfied owners and a competent contractor pool is dependent on accurate adjustments to your production system.

The other aspect of this phase is the dissolution of bonds of dependency between owners and Rehab Specialists. The Construction Management task is complete; now it's time for the owner to begin the task of maintaining the repaired home.

Step 34: Fill Out Evaluations

The use of standardized evaluation forms and procedures is designed to collect data on contractor and program performance. This becomes the data the program uses to weed out incompetent or unscrupulous contractors and improve overall program performance.

Work Tasks

34.1. **Fill out and distribute standardized evaluation forms** (see Addendum 29c, "Procedures to Evaluate Contractors," and Examples 34, "Contractor Evaluation by Rehab Specialist," and 35, "Contractor Evaluation by Homeowner,") assessing the contractor's and program's performance during the execution of the contract. Each person able to contribute to a realistic appraisal of the process should complete the forms. Mail evaluations to the owner and the contractor, and to facilitate return, include stamped, pre-addressed envelopes.

Rehab Specialist

34.2. **Review completed evaluations** for useful suggestions or feedback, recurring problems or bottlenecks. Note for further investigation.

Rehab Specialist

34.3. **File copies of the completed forms in** the Contractor's File, the job file and the program evaluation file.

Rehab Specialist

34.4. **Submit copies of the evaluation** to program management if the evaluation indicates unusually high or low ratings. Investigate low evaluations and use excellent evaluations to identify support and help market the program.

Rehab Specialist

Other Considerations

- Contractor and program evaluation is an ongoing process during construction, and the results of a final evaluation should not be a surprise to anyone.

- Use evaluations to help make changes to the process at least at yearly intervals.

EXAMPLE 34: CONTRACTOR EVALUATION BY REHAB SPECIALIST

NUMBER		YES	NO
1.	Did the work progress without unreasonable or unexplained delays?	✓	
2.	Was there excessive inconvenience for the homeowner?	✓	
3.	Was the contractor available during the job?	✓	
4.	Was the contractor receptive to program concerns (e.g., program image, working relationship with homeowner, organization and procedures)?	✓	
5.	Was the contractor willing to adjust to job changes and problems?	✓	
6.	Did the contractor follow the specifications?	✓	
7.	Were technical guidelines (materials, quality, methods) followed?	✓	
8.	Were attempts made by the contractor to deviate from agreed-upon procedures or work?	✓	
9.	Did the contractor respond to problems and complaints adequately?	✓	
10.	Were appropriate behavior, attitude, and working relationships maintained by the contractor?	✓	
11.	Should this contractor be considered for additional jobs?	✓	

PUT REMARKS ON ANY ITEMS WITH NEGATIVE EVALUATIONS HERE:

8. Change Orders suggested to Homeowner in 3 areas after bid opening

EXAMPLE 35: CONTRACTOR EVALUATION BY HOMEOWNER

NUMBER		YES	NO
1.	Did the work progress without numerous delays?	✓	___
2.	Were you inconvenienced beyond what you were told to expect?	___	✓
3.	Was the contractor available to discuss any questions you may have had?	✓	___
4.	Were the workers courteous and professional?	✓	___
5.	Did the contractor attempt to sell you additional work? Explain: ______________________________	___	✓
6.	Did the contractor suggest changes as to how the work should be done? Explain: Tub surround needed	✓	___
7.	Was the completed work what you expected?	✓	___
8.	Would you recommend this contractor to another homeowner?	✓	___
9.	Did the contractor respond to problems and complaints adequately?	✓	___
10.	Did the contractor explain all procedures and warranties adequately?	✓	___

11. Please rate the quality of the work completed (Circle one)

Very Satisfactory (circled)	Satisfactory	Unsatisfactory	Very Unsatisfactory

EXPLAIN NEGATIVE EVALUATIONS BELOW: ______________________________

Step 35: Perform Warranty Inspections

Warranty inspections are conducted to verify that the completed work is still operating properly.

Work Tasks

35.1. **Schedule warranty inspections** 10 months following the retainage release inspection.

Rehab Specialist

35.2. **Verify appointments** with owners, occupants, and the contractor 10 days prior to a scheduled warranty inspection.

Rehab Specialist

35.3. **Perform the inspection** using the initial set of specifications with change orders. Inspect the property, location by location. Develop a punch list of all unacceptable equipment, or assemblies with moving parts that are defective.

Rehab Specialist

35.4. **Receive owner approval** for the recommended punch list. Using a form letter with the punch list signed by the owner and program director, notify the contractor of the items to be corrected within a 30-day period.

Rehab Specialist

35.5. **Verify completion** of the repairs to the warranty items at the end of the 30-day period by performing a warranty inspection. Log the inspection and any punch lists in the Job File as the final entries and close the file.

Rehab Specialist

Execution Variables

- Omit this step if you want to lower your program overhead and live with a little less quality control. You will be giving up some good information about the contractor's work. If your construction inspections have been vigorous, this should be less of a problem.

ADDENDA

28 Sample Contractor's Warranty
29a Contractor Evaluation by Rehab Specialist
29b Contractor Evaluation by Homeowner
29c Precedures to Evaluate Contractors
30 Automating Specification Writing and Estimating for Housing Rehabilitation

ADDENDUM 1a: SAMPLE HOUSING SURVEY FORM

1st try day/time__________________ Go back time/day__________________

HOUSEHOLD SURVEY

Interviewer(s)______________________________ Date Complete:______________

Dwelling units:______ Address:____

FIRST, I WOULD LIKE TO ASK SOME QUESTIONS ABOUT THE HOME YOU LIVE IN. YOUR ANSWERS WILL BE CONFIDENTIAL AND USED TO PLAN HOW BEST TO HELP YOUR NEIGHBORHOOD WITH HOUSING.

A. Do you own, rent, or have some other arrangement?

1 Own **3** Other (Specify): __________________________
2 Rent or lease **99** NR/DK (No Response/ Don't Know) __________
(1)

B. How much do you spend each month for housing?

RENTER

Rent per month

1	Under $50	**5**	$201 to $250
2	$51 to $100	**6**	$251 to $300
3	$101 to $150	**7**	$301 to $400
4	$151 to $200	**8**	$401 and up

- OR -

HOMEOWNER

House Payment (including taxes, or add in an estimate of taxes)

9	Under $50	**14**	$251 to $300
10	$51 to $100	**15**	$301 to $400
11	$101 to $150	**16**	$401 to $500
12	$151 to $200	**17**	$501 and up
13	$201 to $250		

- OR -

Renter/Owner

18 Nothing
99 NR (No Response) __________
(2)

C. How much per month do you spend in the spring and summer for utilities? (If multi-unit, divide cost)

1	Under $20	7	$121 to $140
2	$ 21 to $ 40	8	$141 to $160
3	$ 41 to $ 60	9	$160 & up
4	$ 61 to $ 80	10	Nothing
5	$ 81 to $100	99	NR (No Response)
6	$101 to $120		

__________ (3)

D. How much do you spend per month in the fall and winter for utilities (Divide cost for multi-unit).

1	Under $20	7	$121 to $140
2	$ 21 to $ 40	8	$141 to $160
3	$ 41 to $ 60	9	$160 & up
4	$ 61 to $ 80	10	Nothing
5	$ 81 to $100	99	NR (No Response)
6	$101 to $120		

__________ (4)

E. OWNERS ONLY: How much do you pay each year for fire insurance? (If multi-unit, divide cost per unit).

1	$ 1 - $100	6	Can't get
2	$101 - $200	7	Don't know
3	$201 - $300	8	Don't want it
4	$301 - $400	99	NR (No Response)
5	$401 - $500		

__________ (5)

F. What don't you like about your house/(apartment)?

(DON'T PROMPT, SIMPLY CIRCLE THE 3 RESPONSES WHICH ARE CLOSEST TO ANSWERS GIVEN TO THIS QUESTION).

1	Too small	9	Bad wiring
2	Too big	10	Needs new kitchen
3	Needs paint	11	Needs new bathroom
4	Needs minor repair	12	Bad porches/sheds
5	Landlord slow to fix	13	Needs roof
6	Can't afford	14	Bad windows
7	Bad neighborhood	15	No heat/poor heat
8	Bad plumbing	99	NR (No Response)

16 Other (specify): __________________________

____

__________ (6)

__________ (7)

__________ (8)

G. RENTERS ONLY: Would you like to buy this house or another house if it were fixed up the way you wanted it and you could afford it?

1	Yes	3	Not sure	
2	No	99	NR (No Response)	________
				(9)

Comments, if any: ____________________

H. RENTERS ONLY: IF YES TO QUESTION G (ABOVE) How much per month would you be willing to pay to own a house that you liked?

2	\$100 - \$150	6	\$300 - \$350	
3	\$150 - \$200	7	\$350 - and up	
4	\$200 - \$250	99	NR (No Response)	
5	\$250 - \$300			________
				(10)

I. OWNERS ONLY: ONLY IF THERE WERE COMPLAINTS ABOUT THE CONDITION OF THE HOUSE (See question F). What help do you need to get your house fixed up? (DON'T PROMPT - PICK UP TO 3)

1	No help	6	Family help	________
2	Loan	7	Grant	(11)
3	Good contractor(s)	8	Time	
4	More income/cash	99	NR (No Response)	________
5	Materials			(12)
9	Other (specify): ____________________			
	____________________			________
				(13)

[THE FOLLOWING QUESTIONS ARE FOR BOTH RENTERS AND OWNERS]

J. How much more would you be willing to pay each month if this place were fixed up?

1	Nothing more	7	\$101 to \$140	
2	Won't take loan	8	\$141 to \$200	
3	\$1 to \$20	9	\$201 and up	
4	\$21 to \$40	10	Not sure	
5	\$41 to \$60	99	NR (No Response)	
6	\$61 to \$100			________
				(14)

K. ONLY IF THERE WERE COMPLAINTS ABOUT THE CONDITION OF THE HOUSE OR APT. (see Question F) Would you or your family be willing to do some of the repairs on your home (apartment) if materials were provided? ________ (15)

1 No 2 Yes 99 NR (No Response)

If so, what kind of repairs? ____________________
(Don't Prompt - Pick up to 2)

3	Painting inside	7	Electrical
4	Painting outside	8	Roofing
5	Carpentry	9	Demolition
6	Plumbing	99	NR (No Response)
10	Other (Specify):________________		

________ (16)
________ (17)

NOW I'D LIKE TO ASK YOU A COUPLE OF QUESTIONS ABOUT YOUR NEIGHBORHOOD

L. Is there an active neighborhood association here?

1	Yes	3	Don't know
2	No	99	NR (No Response)

________ (18)

M. Would you be willing to work with that group or another group to improve housing in the neighborhood?

1	Yes	3	Don't know
2	No	99	NR (No Response)

________ (19)

DEMOGRAPHICS

SO THAT I CAN COMPARE THE NEEDS OF DIFFERENT FAMILIES, I'D LIKE TO ASK ABOUT YOU AND YOUR FAMILY. LET ME REMIND YOU THAT ALL ANSWERS YOU GIVE WILL BE KEPT STRICTLY CONFIDENTIAL.

N. How may years have you lived in your neighborhood?

1	5 or less	4	21 to 30
2	6 to 10	5	31 or more
3	11 to 20	99	NR (No Response)

____
(20)

O. How long have you lived at your current Address?

1	Less than 1 year	5	10 - 20 years
2	1 - 2 years	6	Over 20
3	2 - 5 years	99	NR (No Response)
4	5 - 10 years		

____
(21)

P. Including yourself, how may persons live in your household?

1	One	5	Five
2	Two	6	Six
3	Three	7	Seven
4	Four	8	Eight or more
		99	NR (No Response)

____
(22)

Q. How many elderly persons (62 years or older) live in your household? (If living alone: are you 62 years or older?)

1	One	6	Six
2	Two	7	Seven
3	Three	8	Eight or more
4	Four	99	NR (No Response)
5	Five	0	None

____
(23)

R. How many children 18 years of age or younger live in your household?

1	One	5	Five
2	Two	6	Six or more
3	Three	99	NR (No Response)
4	Four	0	None

____
(24a)

S. How many of these are six years or younger?

1	One	4	Four	
2	Two	5	Five or More	________
3	Three	99	NR ((No Response)	(24b)

T. Have any children been diagnosed to have lead poisoning?

1	Yes	3	Don't Know	________
2	No	99	NR (No Response)	(24c)

U. If yes, has this house been abated for lead?

1	Yes	3	Don't Know	________
2	No	99	NR (No Response)	(24d)

V What is the race of most of the people in your household?(25)

1	White	5	Native American Indian	
2	Black	6	Other (Specify) ________	
3	Hispanic	99	NR (No Response)	
4	Asian			________
				(25)

W. I am going to read several categories of yearly income. When I come to the category that you think includes the total income from everyone in the household per year, please tell me. This information is completely confidential and is for statistical comparisons only. I need only an estimate, not an exact figure.

1	Below $4,500	
2	$4,501 - $7,999	
3	$8,000 - $9,999	
4	$10,000 - $14,999	
5	$15,000 - $19,999	
6	$20,000 - $29,000	
7	$30,000 and above	
8	Not known	
99	NR (No Response)	________
		(26a)

X. What is the source of your income?

1. DSS (AFDC, GA)
2. Social Security
3. Employment
4. Other
5. Pension

________ (26b)

Y. Is there anyone in the household who needs help in obtaining a job?

1	Yes	3	Don't Know
2	No	99	NR (No Response)

________ (27)

Z. Is there anyone in the household who is looking to upgrade his or her reading, education, job training skills?

1	Yes	3	Don't Know
2	No	99	NR (No Response)

________ (28)

ADDENDUM 1b: SHORT FORM HOUSING SURVEY

HOUSING CONDITION

HOUSE TYPE (Circle answer):

1. Number of stories: 1 2 3 4
2. Unit Type: _____Inside _____End
3. Number of bedrooms: 1 2 3 4 5
4. Number of baths: 1 1-1/2 2
5. Does the roof leak? YES NO
6. Do you have good water pressure? YES NO
7. What type of heat? OIL
   GAS
   ELECTRIC
   CENTRAL DUCTED TO ALL ROOMS
   CENTRAL PARTIAL
   SPACE 1,2,3,4
8. How many electric outlets per room average? 1 2 3 4
9. Do the lights have wall switches? All None Some
10. Do you have rats or mice? YES NO
11. Does anything need to be fixed? ______________________
12. ____
13. ____

INTERIOR (Inspect and rate A-D)

14. Wall finish (plaster repair): OK A B C D
15. Floor finish/VCT: OK A B C D
16. Kitchen (42" Cabinets up and down): OK REPLACE
17. Bath (3 pieces working): OK REPLACE

TOILET:	OK	REPLACE
TUB:	OK	REPLACE
SINK:	OK	REPLACE

EXTERIOR (Inspect and rate)

18.	House needs trim painted:	YES		NO		
19.	Entire front painted:	YES		NO		
20.	Conditions of window:	GOOD		OK		WORST
		1	2	3	4	5
21.	Storm windows?	YES		NO		
22.	Needs replacement doors:	YES		NO		
		How many?		1	2	3 4
23.	Any Additions:	YES		NO		
	Condition:	GOOD		OK		WORST
		1	2	3	4	5

HOMEOWNER REQUESTS

1. What items would you like to have done if loan money were available?	How much money would you spend on each of these?
A.______________________	$______________________
B.______________________	$______________________
C.______________________	$______________________
D.______________________	$______________________
E.______________________	$______________________
F.______________________	$______________________
G.______________________	$______________________

ADDENDUM 2: TYPICAL SURVEY REPORT FORMATS

The format you use to present the data you have collected and analyzed will greatly effect its impact. Given the demand characteristics of your situation (i.e., time, budget, availability of expertise, audience sophistication, etc.) the data should be presented in as simple a manner as possible. Three sample formats are presented in this Addendum.

FORMAT A: GENERAL DATA LISTING
Specific Examples of Data Gathered by Using the: SAMPLE HOUSING SURVEY FORM

QUESTION	SURVEY DATA
A.	The majority (59%) owned their own property.
B.	Typically rent was between $100 and $150 per month. The typical average house payment was between $151 and $200 per month.
C.	Summer utilities bills ranged between $41 and $60 per month. However,the modal category was $21 to $40.
D.	Utility costs are higher in winter, with the modal category being $160 and up, or approximately four times the summer figure.
E.	Few people pay less than $100 or more than $300 for fire insurance. The average category was $201 to $300.
F.	Residents' complaints fell mainly in the "other" category, followed by paint, plumbing, and porches.
G/H.	Sixty percent of the renters said they would buy rehabilitated property. However, 73% of the total can only pay less than $100 additional per month.
I.	Twenty-eight percent of owners said they did not need help to fix up their property. Twenty-two percent needed more income. Nineteen percent needed a loan.
J.	Forty-nine percent of residents said they couldn't pay any extra per month to have their place fixed up. Forty-two percent could pay not more than $100 extra.
K.	Thirty-one pecent of respondents would not be able to assist in repairs to their property. Forty-six percent could do interior painting.
L.	Twenty-six percent of respondents are aware of an active neighborhood association in their area. Forty-two percent said there is no active neighborhood association.
M.	Seventy-three percent said they were willing to work with the neighborhood group.

QUESTION	DATA
N.	The majority (76%) lived in their neighborhood more than 5 years. The average category was 11 to 20 years.
O.	Sixty-two percent were at their current address more than 5 years.
P.	Thirty-two percent were living alone and twenty-seven percent were living with only one other person.
Q.	Fifty-five percent of the households have at least one elderly person at that residence.
R.	Sixty-four percent of the households have no children under 18 years.
S.	Fifty-nine percent of the households were black.
T.	Fifty-five percent of the households made less than $10,000 per year household income. Twenty-three percent of the households made less than $5,000 per year.

SAMPLES OF DATA FROM OTHER SOURCES

- Fifty-four percent of the properties are owner occupied, and thirty four percent of the properties with owners lived within the city limits.
- Eighty-two percent of the propeties have land valued at less than $3,000.
- Fifty-seven percent of the properties were in R-2 (duplex) zones.
- Thirty percent were in R-3 (apartment) zones; while seventy-six percent of the structures themselves were single-family design.
- Eighty-seven percent of the properties were approximately 2,500 feet from industrial activities.
- Average rooms per unit are 4, and the mode is 5-room units.
- Ninety-four percent of the properties are one-story houses.
- Forty-two percent of the properties have BCW type foundation.
- There was an average 1,416 square feet per structure.

- The average dwelling unit size was 1,167 square feet.

- The average of the replacement costs as estimated by the Tax Assessment Office was $28,167.

- The cost per dwelling unit was $22,972.

- The average replacement cost per square foot was $19.71.

- The average of the condition percentage as estimated by the Tax Assessment Office was 52%.

- Fifty-seven percent of the houses were built before 1941.

- The average total appraisal market value was $17,981 at a condition level of 52% good.

FORMAT B: DATA PRESENTED BY TOPIC AREA

Organizing data by topic area can help focus the audience on issues that are of importance. The choice of topics should reflect the needs of both the audience and the program.

REHAB COST BY HOUSEHOLD INCOME

1. Forty-six percent of the respondents with a household income below $4,999 had a rehab cost of $1,001-$5,000.

2. All the properties with a rehab cost of $15,001-$20,000 were occupied by those with a household income less than $10,000.

3. All the properties with a rehab cost of $20,001-$25,000 were occupied by those with a household income of $10,000-$14,999.

4. Less than 1% of the properties had a rehab cost over $25,000.

5. Thirty-seven of the properties had a rehab cost of $1,001-$5,000. The average rehab cost was $5,758.65.

6. Twenty-nine percent of the houses with a rehab cost of $1,001-$5,000 were occupied by those with a household income below $4,999, and twenty-seven percent by those with income of $20,000-$29,999.

FORMAT C: SAMPLE DATA CHARTS

Selectively presenting data via charts can be very effective. However, the overuse of charts can obscure important information and confuse the audience. Mixing charts with selected data can enhance the effectiveness of both.

1. The most complaints (23%) of the renters were about paint, followed by plumbing, while the most complaints by owners were "others," followed by porches and plumbing. See the chart below for a detailed breakout.

CHART #1: <u>TYPE OF COMPLAINT BY OWNERS AND RENTERS</u>

CELL CONTENTS: NUMBER OF RESPONSES / PERCENT OF RESPONSES	OWNERS	RENTERS	TOTALS
TOO SMALL	4 7.69	2 5.71	6
NEEDS PAINT	5 9.62	8 22.86	13
NEEDS MINOR REPAIR	1 1.92	2 5.71	3
OWNER SLOW TO FIX	0 0	2 5.71	2
PLUMBING	6 11.54	7 20	13
WIRING	5 9.52	0 0	5
KITCHEN	3 5.77	1 2.85	4
BATHROOM	3 5.77	3 8.57	6
PORCHES	8 15.38	2 5.71	10
ROOF	2 3.85	1 2.86	3
WINDOWS	2 3.85	1 2.86	3
HEATING	1 1.92	3 8.57	4
OTHER	12 23.08	3 8.57	15
TOTALS	52 59.77	35 40.23	87

ADDENDUM 2

CHART #2: OWNER COMPLAINTS BY TYPE OF HELP NEEDED
(CELL CONTENTS, NUMBER OF RESPONSES, COLUMN PERCENT)

	NO HELP	LOAN	GOOD CONTRACTOR	MORE INCOME	MATERIALS	FAMILY HELP	GRANT	OTHER	TOTALS
TOO SMALL	0	0	1	2	0	0	0	0	3
	0	0	50	18.18	0	0	0	0	
NEEDS PAINT	2	0	0	1	1	0	1	0	5
	18.18	0	0	9.09	25	0	12.5	0	
NEEDS MINOR REPAIR	1	0	0	0	0	0	0	0	
	9.09	0	0	0	0	0	0		
PLUMBING	0	1	0	2	1	0	1	1	5
	0	11.11	0	18.19	25	0	12.5	100	
WIRING	0	2	0	1	0	0	1	0	4
	0	22.22	0	9.09	0	0	12.5	0	
KITCHEN	0	1	0	1	0	0	1	0	3
	0	11.11	0	9.09	0	0	12.5	0	
BATHROOM	0	1	0	0	1	0	0	0	2
	0	11.11	0	0	25	0	0	0	
PORCHES	3	0	1	2	1	0	0	0	7
	27.27	0	50	18.18	25	0	0	0	
ROOF	1	0	0	1	0	0	0	0	2
	9.09	0	0	9.09	0	0	0	0	
WINDOWS	0	1	0	0	0	0	1	0	2
	0	11.11	0	0	0	0	12.5	0	
HEATING	0	0	0	0	0	0	1	0	1
	0	0	0	0	0	0	12.5	0	
OTHER	4	3	0	1	0	1	2	0	11
	36.36	33.33	0	9.09	0	100	25	0	
TOTALS	**11**	**9**	**2**	**11**	**4**	**1**	**8**	**1**	**47**

C. WILLINGNESS TO PAY EXTRA FOR A REHABBED HOUSE

1. Fifty-four percent of those who said they could not pay any extra per month if their places were fixed up were owners.

2. Fifty percent of the owners said they could not pay any extra per month if their places were fixed up, while forty-eight percent of the renters said they could not pay any extra.

3. Those willing to pay more than $100 extra per month for their places to be fixed up were all owners.

CHART #3: <u>WILLINGNESS TO PAY EXTRA</u>

CELL CONTENTS:
NUMBER OF RESPONSES
COLUMN PERCENT

	OWNERS	RENTERS	TOTALS
NOTHING MORE	21 50	18 48.65	39
WON'T TAKE LOAN	2 4.76	0 0	2
$1-20	3 7.14	7 18.92	10
$21-40	2 4.76	2 5.41	4
$41-60	4	3	7
$61-100	5 14.29	4 10.81	10
$101-140	1 2.38	1 0	1
$141-200	2 4.76	0 0	2
NOT SURE	1 2.38	3 8.11	4
TOTALS	42	37	79

ADDENDUM 3: TYPICAL PROGRAM CONSTRUCTION STANDARDS

Purpose

This is a minimum standard for all single-family housing in our service area. Where it is economically feasible, we want all housing to be:

1) Reasonably free from fire and other hazards.

2) Comfortable, healthy and well-ventilated.

3) Decent appearing inside and out.

4) Inexpensive to heat and maintain.

5) Solid structurally and weathertight.

6) Reasonably equipped for current handicapped occupants.

Because financial resources for rehabilitation will be hard to obtain for the neediest families, this standard does not require such work as converting safe space heat to central heat, installing air conditioning, "modernizing" functional kitchens and bathrooms, or doing purely cosmetic treatments to exteriors, floors, windows, etc. This standard does not apply to emergency repairs or weatherization work.

Compliance with Codes and Other Standards

This standard equals or exceeds the Housing Quality Standards of the U.S. Department of Housing and Urban Development.

This standard meets the requirements of the local building, plumbing and electrical codes as they pertain to existing housing, but not as they pertain to new construction.

When a single room has its interior finish (plaster or wallboard) removed, the rehabilitation in that single room will meet new construction standards for plumbing, electricity, heating and fire ratings. When such removal occurs in over 50% of the rooms in a freestanding building, this rule will apply to the whole building. Otherwise, the code for existing housing will prevail.

Local and federal regulations pertaining to zoning, traffic, drainage, flood-plains and fire limits will be observed.

<u>Exceptions</u>

It is the purpose of this organization to bring all housing in its area to this standard except when the rehabilitation is not economically feasible -- that is, the existing debt on the property plus the program loan exceeds ___% of after-rehabilitation market value or $______ (whichever is greater).

<u>Practical Effect</u>

This organization will make rehabilitation grants and/or loans to qualified families with incomes under $_________ for improvements in conformance with this standard. Unless another standard or policy statement of this organization permits it, no work items may be undertaken during a rehabilitation project that exceeds this standard. If proposed improvements meet a lesser standard (e.g., emergency repairs, weatherization or "staged" improvements), they will conform to standards for that program or standards negotiated on a case-by-case basis by this agency and local code authorities.

<u>Exterior and Structural Standards</u>

<u>Dilapidated outbuildings, fences and other structures</u> - Unsafe structures and attractive nuisances will be demolished and removed. Visually blighted structures will be demolished and removed unless their use is essential to the lifestyle of occupants.

<u>Debris, junk and unwanted trees</u> - Debris, junk and dead or dangerous tree growth will be removed.

<u>Repair of outbuildings</u> - Unsafe, unsound or unsightly outbuildings and other structures (fences, walks, etc.) will be repaired one time up to a total estimated cost per dwelling unit of $1,500.

<u>Walks and driveways</u> - Badly deteriorated essential paving will be repaired or replaced to result in at least a 7-year useful life after rehab. Nonessential deteriorated paving will be removed.

<u>Exterior requirements</u> - Every dwelling unit will have a mailbox or mail slot and 3" high address numbers.

<u>Steps, stairways and porch decks</u> - These will be reasonably level, even surfaces and have an expected useful life of 5 years or more after rehab.

<u>Porches</u> - Unsound, unsafe or unsightly porches that are unwanted or infeasible to repair will be removed. Porch repairs will be made up to a total estimated cost of $1,500 per dwelling unit to result in an expected useful life of 10 years.

Handrails and guardrails - These are required on one side of steps or stairs with more than two risers, and around porches or platforms over 30" high. Portions of stairs or platforms over 42" above grade or above another floor will have guardrails or balustrades.

Siding and trim - These will be intact and weatherproof. No component will have an expected useful life less than 10 years. All components will have a continuous coat of paint or bonded finish with an expected life of at least 5 years. Siding requiring over 40% replacement will be re-sided entirely with vinyl or aluminum.

Foundations - These will be sound, reasonably level and reasonably free from movement for an expected period of at least 15 years.

Structural members - Structural framing and masonry should appear to be free from deterioration, rot or serious termite damage, be adequately sized for current loads and have a 15-year expected useful life after rehab. Prior to any rehab, all sagging floor joists or rafters will be visually inspected. Significant structural damage and its cause will be corrected.

Insulation - Attic areas and floors over cold crawl spaces will be insulated to R-21 or to cavity capacity if less than 3 inches of insulation is currently in place. Walls will be insulated only if the plaster or wallboard is removed. Plastic ground covers will be placed in crawl spaces under insulated floors. Recessed ceiling light fixtures will be covered, leaving an air space, before insulation is placed over them.

Attic and crawl space ventilation - Attics and crawl spaces will be ventilated to remove excess moisture at a minimum ratio of 1 square inch of vent for each 300 square inches of floor.

Exterior doors - Doors and hardware must have at least a 5-year useful life after rehab. All hollow-core exterior doors will be replaced. Locks will tightly secure doors. If desired by occupants, and required to meet ventilation standards, screen doors will be installed on one or two doors per dwelling unit.

Roofs - Roof covering will be replaced if blistered and the mineral covering is substantially deteriorated. Missing or leaking shingles and flashing on functional roofs will be repaired. Broken antennas will be removed. Roofing material shall have at least a 7-year expected useful life after rehab (assuming normal maintenance), or shall be replaced with new roofing.

Chimneys - Unsound chimney tops will be repaired or removed. When chimneys are being used for venting, they will be restored, if necessary, to safe and operable condition, with an expected useful life of at least 15 years.

Heating System Standards

Heating systems - Inoperative, hazardous or inefficient (under 60% efficiency per a test) heating systems will be repaired to perform at 78% or better efficiency, or replaced. Replacement space heating systems are acceptable where they are preferred by the occupants and have a lower rehabilitation cost and the same or lower operating cost as compared to central heat. All heating equipment will have a minimum expected useful life of 10 years, if pre-existing, or 20 years, if new. Fuel tanks will be located in fire resistant areas. Habitable rooms must be provided with 68° heat when the outside temperature is at the average yearly minimum for the locale.

Wood and Coal Stoves - Wood and coal stove installations will have safe clearances from combustible surfaces as defined by the local Fire Marshal. Protective floor and wall coverings will be provided as necessary. Flues will be safe, properly sized and clean.

Electrical System Standards

Electrical outlets - Every room will have at least one duplex receptacle and one light fixture or two duplex recepticals. Large rooms or those with heavy current electrical loads may require three or more outlets, as determined by the program inspector. Kitchens will have at least two outlets on separate circuits, one of which must have 20 amperes capacity, and one of which may be 15 amperes. Bathrooms will have at least one grounded outlet.

Electrical wiring - "Knob-and-tube" and other older-type wiring will be left in place if safe and servicable. Where areas are being insulated, knob-and-tube will be removed and all junction boxes covered. Recessed ceiling light fixtures will be covered before insulation is placed over them.

Ground fault interrupters - Electrical outlets within 4 feet of a bathtub must be removed, or replaced with a GFI-protected outlet. Any newly wired outlet in a bathroom or within 5 feet of a kitchen sink area must be GFI-protected.

Light fixtures - Every habitable room must be provided with a light fixture or one outlet controlled by a wall switch. Wall or ceiling lights with their own internal switches are acceptable. Common or internal stairways from one floor used as a living area to another must have at least one light controlled by a 3-way switch, or in the case of public hallways, a 3-way switch or automatic switching device. Exterior stairs over five risers will have a light fixture.

Appliance outlets - Safe electrical outlets must be provided for all existing or appliances or those proposed to be installed during the rehab. Appliances requiring 120 volts may be on a circuit with other outlets so long as there is no obvious evidence the circuit is overloaded. Only essential motors requiring 240 volts must have a separate, dedicated circuit and outlet.

Electrical service entrance - The service entrance (whether new or pre-existing) shall be in good condition and sized to safely service all outlets, fixtures and appliances proposed to remain or be newly installed during the rehab. In no case will it be less than 60 amperes. Any service entrance replacement will have a minimum capacity of 100 amperes.

Smoke Detectors - A smoke detector shall be required on each floor that contains a bedroom.

Plumbing System Standards

Kitchen plumbing - Every dwelling unit will have at least a single-bowl kitchen sink with hot and cold water located in the kitchen.

Bathroom plumbing - Every dwelling unit will have an attached bathroom containing a lavatory (basin), where space permits, commode and shower or tub.

Plumbing fixtures - The essential fixtures named above will be in good repair with an expected useful life of 10 years after rehab. They will be replaced only when repair costs exceed 50% of new fixture costs. Faucets will be free from leaking for an estimated 5 years after rehab.

Water shutoffs - Malfunctioning existing shutoffs will be repaired or removed. There will be at least one accessible main shutoff interior to the structure for each dwelling unit after rehab.

Water supply lines - All hot and cold water lines will be expected to have a 15-year useful life after rehab, without major leaks, and maintain a 3 gallon per minute flow. Pressure regulators will be installed where street pressures may exceed 100 p.s.i. Exposed piping, PVC, and polypropylene are allowed.

Water waste lines - Existing drain lines must safely and adequately carry away waste water, and must be vented well enough to prevent significant breaks of the water seal in fixture traps. Existing S-traps or drum traps may remain. Existing traps that are broken may be replaced with similar devices unless the entire waste water system is being replaced in that room, in which case the requirements for new construction will govern. Exposed piping, PVC and ABS are allowed. Minimum expected life is 15 years after rehab.

Wells - Every house not connected to a public water system will have its own system offering safe water, with at least 20 pounds pressure at each faucet.

Water heaters - Each dwelling unit shall have a minimum of one 30-gallon hot water heater (or a shared water heater with similar capacity) in good working condition with an expected life of 5 years. Durable insulating jackets will be provided if this does not require relocating the water heater and results in a safe condition. Water heaters should produce at least 120°F water at all taps. Relief valves and overflow pipes will be properly installed.

New plumbing system - Where plumbing lines and fixtures are being torn out in a room or entire dwelling unit, new construction codes will be observed to the extent practicable for that room or dwelling unit.

Showers - This standard does not require the installation of a shower where a workable tub with no shower exists.

Shower areas - Showers, where in operation, must have a watertight wall with an expected useful life of at least 5 years.

Other Interior Standards

Floors - Bathroom and kitchen floors will have a water-resistant, easily cleanable surface. Damaged wood floors will be repaired and painted. Seriously damaged tile or vinyl flooring will be replaced. Basement floors shall be a minimum of 3" poured concrete.

Insect screens - Every habitable room having windows and/or doors will have at least one window or door screen.

Bathroom ventilation - Natural or mechanical ventilation is required.

Kitchen ventilation - An openable window or vent fan must be present in working order.

Kitchen cabinets - 1 and 2 bedroom dwelling units must have one base and one wall cabinet at least 42 inches long. 3+ bedroom units must have 72" base and 48" wall cabinets.

Storage areas - A built-in storage device at least two feet wide with a shelf and clothes rod are required in or near each bedroom.

Walls - Walls and woodwork should be free from peeling paint and loose or cracked plaster. Rooms with deteriorated or stained paint or wallpaper will be repainted. Solid plaster or wall board that has uneven but sound surfaces will be left alone.

Interior doors - Any doors that are damaged and beyond repair will be replaced. Slide bolts are allowed. All bedrooms and the bathrooms will have operable doors with a latch.

Windows - Each habitable room, excluding the kitchen area or bathroom, will have at least one operable window or skylight, with at least one operable sash with a working device to hold it open. All windows accessible from the ground or porches will have locking hardware.

Storm windows - Where practicable, every single-glazed window will have an outside storm window in good condition.

Weatherstripping - This will be installed in a prudent manner, recognizing the oxygen requirements and pollution production of heating devices that may be in place. Homes that are heated exclusively with electricity or well-vented furnaces or stoves will be tightly weatherstripped.

<u>Room sizes, layouts, fire ratings and fire exits</u> - The dwelling unit, after rehab, will conform to the requirements in these areas of the BOCA Existing Structures Code/1987, unless unusual conditions merit consideration of a waiver.

<u>Hazardous Materials</u> - Asbestos, radon and lead paint, when discovered, shall be abated in conformance with applicable local, state and federal law.

<u>REPLACEMENT STANDARDS</u>

<u>Replacement Standards</u> - Used, discontinued and cosmetically blemished materials are allowed when they meet the performance standard and are owner approved.

The following are not performance standards, but standards for building components replacement are specified.

Electric: 100 AMP main disconnect panel with room for 12 circuits

Plumbing: Copper supply; 3/4" copper main supply from street
30 gal. gas or 40 gal. electric water heater with 7 year warranty.

H.V.A.C.: Gas or Oil Furnace or boiler with A.F.U.E. rating of 78 or better, exposed duct work painted

Roofing: Three-ply built up or double-coverage shelvage with 10-year guarantee

Siding: 210-lb class A fiberglas shingles or stained T-1-11 over frame walls

Window Sash: Aluminum, single-glazed one over one or to match existing configuration

ADDENDUM 4: REHAB LOAN APPLICATION

Owner ______________________ Date ______________________

Co-owner ______________________ Telephone ______________________

Address ______________________

____

HOUSEHOLD MEMBERS:

Name	Age	Relationship	Monthly Income	Source of Income

A. GROSS MONTHLY INCOME: (Include all of the following:)

- Owner's gross salary ______________
- Co-owner's gross salary ______________
- Other household members' gross salary ______________
- AFDC, SSI or other pensions/annuities ______________
- Disability ______________
- Other income (child support, alimony) ______________
- Gross rental income ______________

TOTAL GROSS FAMILY INCOME: $______________

B. TOTAL DEDUCTIONS FROM GROSS INCOME:

1. Federal Income Tax $______________
2. State Income Tax ______________
3. Social Security, retirement ______________

TOTAL DEDUCTIONS: $______________

C. TOTAL NET INCOME: (A minus B) $______________

D. MONTHLY EXPENSES:

1. All loan payments on house $______________
2. Insurance on house ______________
3. Real estate taxes ______________
4. Electric ______________

5. Both gas $________
6. Wood ________
7. Oil ________
8. Maintenance on house (estimate monthly average) ________

SUBTOTAL OF HOUSING EXPENSES: $________

9. Telephone ________
10. Food (purchased with cash) ________
11. Clothing ________
12. Transportation ________
13. Auto Insurance ________
14. Medical (Including insurance) ________
15. Life Insurance ________
16. Tuition, books ________
17. Loan and credit card payments - (not on house) ________
18. Alimony, child support (paid out) ________
19. Rental property ________
20. Other: ________________ ________

TOTAL HOUSEHOLD EXPENSES: $________

E. TOTAL INCOME AVAILABLE FOR LOAN PAYMENTS: $________
(C minus D)

F. LIABILITIES:

Creditor	Account Number	Date Opened	Present Balance	Monthly Payment	Payments Past Due (#)

G. ASSETS:

1. Bank Accounts:

Checking: ________________
Name

____
Account Number

____
Balance

Savings: ________________
Name

____
Account Number

____
Balance

Current Value

2. U.S. Savings Bond: $__________

3. Marketable Securities: $__________

   Name ________________ $__________

   Name ________________ $__________

4. Residence purchase price $__________
   Estimated market value $__________
   Mortgage amount $__________
   Current balance $__________

5. Other (describe) __________
   Land value (if not included above) $__________
   ________________________ $__________

TOTAL VALUE OF ASSETS: $__________

H. EMPLOYMENT INFORMATION:

1. Person with job (name)____________________ AGE: ____
   Employer Address ________________________
   ____
   Telephone ________________________
   How long? ________________________
   Position ________________________

2. Person with job (name)____________________ AGE: ____
   Employer Address ________________________
   ____
   Telephone ________________________
   How long? ________________________
   Position ________________________

Owner's Signature____________________________

Co-owner's Signature__________________________

____

____

OFFICE USE ONLY:

Estimate of Housing Expenses after Repairs (Monthly)

1.Loan 1:________________________ ____________________________________

2.Loan 2:________________________ ____________________________________

3.Insurance ____________________________________

4.Real Estate Taxes ____________________________________

5.Electric ____________________________________

6.Both Gas ____________________________________

7.Wood ____________________________________

8.Oil ____________________________________

Total Housing Expenses $____________________________________

ADDENDUM 5: INSPECTION CHECKLIST

Project (Month-Year):____

Address of Property:____

Unit Name or Number:____

Inspector's Name:____

Agency Program:____

Building Description

No. of Dwellings Units:____

Type of Construction:____

No. of Bedrooms per Dwelling Unit (Indicate location and number of bedrooms for each dwelling unit) ________________________________

Exterior Dimensions ___________________ by ___________________

Names of Subcontractors for This Job

Heating ____

Plumbing ____

Electrical ____

Other ____

____

____

____

____

General Requirements

1- 4	___________	M	Permits & Fees
1- 52	___________	EA	Final Clean
1- 60	___________	AL	Contingency Pre-Bid $ ________
26-7940	___________	EA	Exterminate Rodents
26-7904	___________	EA	Exterminate Termites
5- 601	___________	LD	Haul Debris to Dump
___-___	___________	__	________________________________
			________________________________

EXTERIOR

ROOF

9- 2350	_______	SQ	Fiberglass Shingle-Roof Over
9- 2356	_______	SQ	Tear off/shingle
9- 2272	_______	SQ	Flat Roof-Recoat
9- 2280	_______	SQ	3-Ply Built-Up
9- 2300	_______	SQ	90 LB Roll Roof
9- 2332	_______	PC	Repair Slate
9- 2258	_______	SQ	Sheathing - 1/2"
9- 2344	_______	SF	Shingle Repair Custom $____________
__- ____	_______		______________________________

CHIMNEY

7- 1154	_______	EA	Tuckpoint Chimney
7- 1160	_______	EA	Chimney Cap
7- 1170	_______	LF	Rebuild Chimney
9- 2505	_______	EA	Chimney -- Reflash

GUTTERS DOWNSPOUTS

31- 2416	_______	LF	New Gutter/Spouts 5"
31- 2494	_______	LF	Gutter/Soffit/Trim
31- 2410	_______	LF	Rehang Gutter/Spout
31- 2422	_______	EA	Splash Block
31- 2438	_______	LF	Gutter/Spout Custom $____________

DOORS & WINDOWS

14- 1930	_______	EA	Remove covering from opening
32- 3340	_______	EA	S.C. Flush Door w/lock
32- 3342	_______	EA	Ext-Panelled Door w/lock
32- 3350	_______	EA	Prehung Metal Entry Door w/lock
14- 3304	_______	EA	Rehang Ext. Door
14- 3310	_______	EA	Jamb Replace -Ext.
14- 3324	_______	EA	Double Key Deadbolt
10- 3334	_______	EA	Weatherstrip Door
14- 4302	_______	EA	Lock - Security Slidebolt
14- 4309	_______	EA	Door Knocker
14- 4350	_______	EA	Mail Slot
14- 4352	_______	EA	Mailbox
14- 4340	_______	EA	House Number Set
23- 7364	_______	EA	Entrance Light
12- 3052	_______	EA	Storm Window
11- 3012	_______	EA	Vinyl DH S.G. Window
14- 2790	_______	EA	Repair Storm Door
12- 2792	_______	EA	Storm Door and Screen
__- ____	_______		______________________________

SIDING

23- 7370	________	EA	Flood Light - Single Bulb
14- 2478	________	LF	Roof Fascia 1x6
14- 2484	________	SF	Soffit - 1/4" Plywood
14- 3104	________	LF	Replace Clapboard
14- 3122	________	SF	Install T-1-11
14- 3130	________	SF	Cedar Shingle Repair
14- 3138	________	LF	1x4 Wood Corner Trim
13- 3164	________	SF	Vinyl-Entire
13- 3170	________	LF	Vinyl-Wrap Trim
13- 3194	________	SF	Alum-Replace
13- 3196	________	SF	Alum-Entire
13- 3210	________	LF	Alum-Wrap Trim
13- 3232	________	SF	Cement Shingles-Repair
13- 3234	________	SF	Asphalt Shingles
__- ____	________	__	______________________________
__- ____	________	__	______________________________

PORCH	**FRONT**	**REAR**	**SIDE**		
5- 710	_____	_____	_____	EA	Demo Small Porch $ ___
14- 1884	_____	_____	_____	SF	Porch -- Rebuild and Paint
6- 890	_____	_____	_____	EA	Repair Concrete Steps
6- 898	_____	_____	_____	RI	Concrete Stoop- Form in Place
14- 1502	_____	_____	_____	EA	Tread Repl. - Ext.
14- 3842	_____	_____	_____	RI	Stairs -- Wood
14- 1540	_____	_____	_____	RI	Steps & Landing-Repl. Ext.
14- 1570	_____	_____	_____	LF	Wood Handrail - Ext.
14- 1802	_____	_____	_____	LF	4x4 Post-Support Roof
6- 804	_____	_____	_____	EA	Footing Pier
14- 1822	_____	_____	_____	LF	Deck Joist 2x6
14- 1824	_____	_____	_____	LF	Deck Joist 2x8
14- 1830	_____	_____	_____	SF	T & G Deck
14- 1860	_____	_____	_____	SF	T & G Porch Ceiling
24- 1850	_____	_____	_____	LF	Wrought Iron Railing
14- 3810	_____	_____	_____	LF	Porch Rail Repair -- Wood
14- 3814	_____	_____	_____	LF	Porch Rail -- Replace -- Wood
14- 1870	_____	_____	_____	SF	Replace Lattice
14- 2478	_____	_____	_____	LF	Fascia 1x6
14- 5352	_____	_____	_____	SF	Paint Porch
14- 4603	_____	_____	_____	EA	Misc. Carpentry 3 $______
__- ____	_____	_____	_____	__	________________________

PAINTING

17- 5282	________	SF	Paint Exterior Wood
17- 5320	________	SF	Paint Exterior Trim
17- ____	________		________________________________

MASONRY

7-	1086	_______	SF	Tuckpoint Foundation
6-	874	_______	EA	Concrete Rat Hole
7-	1094	_______	SF	Brick Wall Repair
7-	1232	_______	SF	Parge/Bond Foundation
7-	1250	_______	EA	Foundation Vent
__-	____	_______		______________________________

YARD AND HAULING

6-	882	_______	SF	Sidewalk Repair
6-	884	_______	SF	Sidewalk - Concrete
30-	310	_______	LF	4' CH Link Fence
30-	360	_______	EA	4' CH Link Gate
5-	772	_______	LD	Clean Yard-Remove Debris
5-	702	_______	EA	Demo Out Building $______________
22-	6580	_______	EA	Hose Bibb
__-	____	_______		______________________________

MISCELLANEOUS

__-	____	_______		______________________________
__-	____	_______		______________________________

BASEMENT ____________ by ___________

ELECTRIC

23-	6980	_______	EA	100 AMP Main 10 Breaker Panel
23-	6912	_______	EA	Relocate Electric Panel
23-	7332	_______	EA	Basement Rewire, Inc. Panel
23-	7096	_______	EA	Porcel. P.C. Fixture
23-	7095	_______	EA	Replace P.C. Fixture
23-	7334	_______	EA	Rewire House to Code
__-	____	_______	EA	Custom Electric $______________
				______________________________

HOT WATER TANKS

22-	6530	_______	EA	40 Gal. Hot Water Tank
22-	6527	_______	EA	Reinstall Flue Pipe
__-	____	_______	EA	$______________________________
22-	____	_______	__	______________________________

WATER LINES AND RISERS

22-	6205	_______	EA	Replumb Basement
22-	6204	_______	LF	Water Service -K Line
22-	6232	_______	LF	Copper Risers and to Lines
22-	6240	_______	EA	Shut Off Valve
__-	____	_______	__	______________________________

DRAIN WASTE AND VENT LINES

22- 6200	_______		General Requirements
22- 6282	_______	LF	Repair Waste Line w/PVC
22- 6290	_______	EA	Snake Waste Line
22- 6300	_______	LF	New DWV Line - PVC
22- 6304	_______	EA	Clean-Out Tee on Waste Line
22- 6330	_______	LF	Sewer Service/4 Inch PVC
22- 6726	_______	EA	Replumb House 1 Bath
__- ____	_______	__	________________________________

LAUNDRY AREA

22- 6554	_______	EA	Single Bowl Laundry Tub
22- 6560	_______	EA	Hose Bibbs for Washer
23- 7020	_______	EA	Laundry Outlet 110/20
23- 7021	_______	EA	Electric Dryer Outlet 220/30
22- 6570	_______	EA	Gas Dryer Hookups
22- 6392	_______	EA	Repair Faucet
22- 6562	_______	EA	Faucet -- Laundry Tub
22- 6278	_______	EA	Custom Plumbing Supply $__________
22- 6358	_______	EA	Custom DWV Lines $______________
22- 6388	_______	EA	Custom Gas Lines $______________
22- 6578	_______	EA	Custom Laundry Hookup $__________

GAS LINES

22- 6362	_______	LF	Gas Line
22- 6374	_______	EA	Gas Shut-Off Valve
22- 6382	_______	LN	Gas Line Removal and Cap
__- ____	_______	________________________________	

FURNACE

21- 5800	_______	__	General Requirements
21- 5806	_______	EA	Test Furnace/Give Opinion
21- 5807	_______	EA	Furnace Repair/Cost Allowance $_____
21- 5826	_______	EA	Gas Furnace - 100,000 BTU
21- 5804	_______	EA	Clean and Adjust Furnace
21- 6140	_______	EA	Thermostat
21- 6152	_______	EA	Flue Thimble
21- 6160	_______	EA	Flue Replace
21- 6104	_______	EA	Cold Air Drop
21- 5898	_______	EA	Furnace - Gas Custom $____________
__- ____	_______	__	____________________________

BOILER

22- 5954	_______	EA	Test Boiler/Give Opinion
22- 5955	_______	EA	Boiler Repair-Cost Allowance
22- 5970	_______	EA	Boiler - Gas Replace
22- 5952	_______	EA	Burner Maintenance
22- 5982	_______	EA	Boiler - Gas Custom $_____________

INTERIOR MASONRY

7- 1232	_______	SF	Parge & Bond Foundation
7- 1154	_______	SF	Tuckpoint Chimney
7- 1086	_______	SF	Tuckpoint Foundation
7- 1020	_______	SF	Block Wall Repair
7- 1094	_______	SF	Brick Wall Repair
6- 872	_______	SF	Concrete Floor Patch
7- 1250	_______	EA	Foundation Vent
__- ____	_______	___	$________________________________

CARPENTRY

8- 1520	_______	RI	Cellar Staircase Replace
8- 3514	_______	EA	Window Repair
8- 3540	_______	EA	Window-Basement Unit
8- 3586	_______	SF	Glass Block
8- 3502	_______	EA	Sash Lock
8- 3574	_______	EA	Glass Replace
8- 1550	_______	EA	Handrail Repair
8- 1561	_______	LF	Handrail 2"x4"
8- 1579	_______	EA	Staircase Repair - Custom $_________
8- 4604	_______	EA	Misc. Carpentry 4 $________________
__- ____	_______	__	______________________________
__- ____	_______	__	______________________________

KITCHEN ____________ by ___________

PLUMBING

22- 6418	_______	EA	Sink-Double Bowl Complete
22- 6392	_______	EA	Faucet Repair
22- 6398	_______	EA	Single Control Faucet
22- 6404	_______	EA	Dual Control Faucet
22- 6240	_______	EA	Shut Off Valve
22- 6242	_______	EA	Supply Line and Shut Off
22- 6362	_______	LF	Gas Line
22- 6374	_______	EA	Gas Shut Off Valve
22- 6368	_______	EA	Gas Flex Supply Line
22- 6279	_______	EA	Plumbing - Custom $________________
__- ____	_______	__	______________________________

ELECTRIC

23- 7004	_______	EA	Replace Wall Receptacle
23- 7019	_______	EA	GFI Countertop
23- 7014	_______	EA	Receptacle -230 AMP Wire
23- 7010	_______	EA	Receptacle - 15 AMP Wire
23- 7072	_______	EA	Replace Switch
23- 7094	_______	EA	Replace Light Fixture
23- 7092	_______	EA	Light Fixture and Switch
23- 7090	_______	EA	Light Fixture Globe
23- 7108	_______	EA	3-Way Switch & Fixture
23- 7066	_______	EA	Electric - Custom $________________
__- ____	_______	__	______________________________

CABINETS AND SHELVING

34- 3602	_______	EA	Cabinets-Repair Base and Wall
34- 3612	_______	PR	Cabs Wall & Base 5' w/D.B. Sink
34- 3614	_______	LF	Wood Base Cabinet
34- 3624	_______	LF	Wood Hanging Cabinet
34- 3640	_______	LF	Countertop
__- ____	_______	__	______________________________

SHEET GOODS AND FLOOR TILE

19- 5520	_______	SF	Vinyl Sheet Goods
19- 5519	_______	RM	Underlay and VCT
19- 5524	_______	SF	Underlay & Vinyl Sheet Goods

HVAC

21- 6172	_______	EA	Floor Register
21- 6102	_______	EA	Heat Duct & Register
21- 6138	_______	EA	Custom HVAC Repair $_____________

WALLS AND CEILINGS

15- 5134	_______	SF	Gut and Drywall - 1/2" W C
15- 5122	_______	SF	Shellcoat Drywall - 3/8th" W C
15- 5162	_______	SF	Drywall-Fur, Hang, Finish
15- 5104	_______	SF	Drywall - Renail and Retape W C
15- 5108	_______	SF	Drywall - Patch W C
15- 5003	_______	RM	Strip Paper, Spackle, Prime W C
15- 5004	_______	SF	Patch Plaster -- Large Area
8- 4002	_______	SF	Ceiling Tile Repair
8- 4290	_______	LF	Shoe Mold
8- 4286	_______	LF	Baseboard - 1x4
8- 4602	_______	EA	Misc. Carpentry 2 $_____________
__- ____	_______	__	______________________________

PAINTING

35- 5255	_______	EA	Paint Vacant Room -- Gloss
35- 5224	_______	SF	Prep & Paint Interior Trim
35- 5230	_______	EA	Prep & Paint Cabinets

WINDOWS

8- 3502	_______	EA	Sash Lock
8- 3504	_______	EA	Sash Lift
8- 3506	_______	PR	Sash Cords
8- 3514	_______	EA	Window Repair
8- 3574	_______	EA	Glass Replace
8- 3534	_______	EA	Sash Replace

DOORS

8- 3410	_______	EA	Rework Int. Door
8- 3402	_______	EA	Door -- Remove
8- 3430	_______	EA	Door -- Flush Interior
__- ____	_______		______________________________

BATHROOM ____________ by ___________

PLUMBING

22- 6240	_______	EA	Shut Off Valve
22- 6242	_______	EA	Supply Line and Shut Off
22- 6279	_______	EA	Plumbing -- Custom $________________
22- 6442	_______	EA	Faucet Repair
22- 6460	_______	EA	Lavatory-Wall Hung Complete
22- 6448	_______	EA	Dual Control Faucet
22- 6450	_______	EA	Secure Wall Hung Lavatory
22- 6466	_______	EA	Vanity 24" Complete
22- 6488	_______	EA	5' Steel Tub w/Shower Complete
8- 4130	_______	EA	Tub end Wall
22- 5422	_______	EA	Tub Surround
22- 6470	_______	EA	Diverter & Shower Ring
22- 4416	_______	EA	Shower Curtain Rod
22- 6486	_______	EA	Water Saver Showerhead
22- 6512	_______	EA	Replace Commode
22- 6506	_______	EA	Repair Commode (Guts)
22- 6712	_______	EA	Lavatory & Commode - All Work
22- 6718	_______	EA	Lav+Commode+Tub/Sh. - All Work
__- ____	_______	__	____________________________________

ELECTRIC

23- 7018	_______	EA	GFI Receptacle
23- 7072	_______	EA	Replace Switch
23- 7094	_______	EA	Replace Light Fixture
23- 7092	_______	EA	New Light Fixture & Switch
23- 7090	_______	EA	Light Fixture Globe
23- 7066	_______	EA	Electric Repair -- Custom $_________
__- ____	_______	__	____________________________________

SHEET GOODS AND FLOOR TILE

19- 5540	_______	EA	Underlay and VCT
19- 5524	_______	SF	Underlay & Vinyl Sheet Goods
19- 5504	_______	SF	Tile flooring - Repair

WALLS AND CEILINGS

15- 5134	_______	SF	Gut and Drywall - 1/2" W C
15- 5122	_______	SF	Shellcoat Drywall - 3/8th" W C
15- 5104	_______	SF	Drywall - Renail and Retape W C
15- 5004	_______	SF	Patch Plaster -- Large Area
8- 4290	_______	LF	Shoe Mold
8- 4286	_______	LF	Baseboard - 1x4
8- 4162	_______	EA	Trim Door Set - 1x4
8- 4166	_______	EA	Trim - Window Set 1x4
__- ____	_______	__	____

CARPENTRY

8- 3502	________	EA	Sash Lock
8- 3506	________	EA	Sash Cords
8- 3514	________	EA	Repair Window
8- 3572	________	EA	Glass Replace
8- 3574	________	EA	Wood Sash - Replace
8- 3534	________	EA	New Window Sash
8- 3544	________	EA	Replace Double Hung Window
8- 3410	________	EA	Rework Interior Door
8- 4376	________	EA	Privacy Lock
8- 3470	________	EA	Door Stop
33- 3430	________	EA	Door -- Flush Interior
8- 4402	________	EA	Towel Bar
8- 4412	________	EA	Bathroom Accessory Set
8- 4408	________	EA	Medicine Cabinet
8- 4601	________	EA	Misc. Carpentry 1 $________________
__- ____	________	__	______________________________

PAINTING

35- 5255	________	EA	Prep & Paint Vacant Room -- Gloss
__ ____	________		______________________________

INTERIOR ROOM ____________________ by __________

LR:________ **DR:** ___________ **HALL:**__________ **BR:**___________ **OTHER:** ________

ELECTRIC

23- 7006	________	EA	Electric Cover Plate
23- 7004	________	EA	Replace Wall Receptacle
23- 7010	________	EA	Receptacle - 15 AMP Wire
23- 7072	________	EA	Replace Switch
23- 7094	________	EA	Replace Light Fixture
23- 7076	________	EA	Switch Light
23- 7080	________	EA	Switch Wall Receptacle
23- 7092	________	EA	Light Fixture and Switch
23- 7090	________	EA	Light Fixture Globe
23- 7108	________	EA	3-way Switch & Fixture
23- 7126	________	EA	Smoke Detector - Hard Wired
23- 7066	________	EA	Electric Repair -- Custom $______________

HVAC

21- 6173	________	EA	Register -- Wall
21- 6172	________	EA	Floor Register
21- 6102	________	EA	Heat Duct & Register
21- 6138	________	EA	Custom HVAC Repair $______________________
21- 6144	________	EA	Thermostat -- Mechanical Setback
__- ____	________	__	______________________________

WINDOWS

8- 3502	_______	EA	Sash Lock
8- 3506	_______	EA	Sash Cords
8- 3504	_______	EA	Sash Lift
8- 3514	_______	EA	Repair Window
8- 3574	_______	EA	Wood Sash - Replace Glass
8- 3534	_______	EA	Window Sash - Replace
8- 3544	_______	EA	Replace Double Hung Window
8- 3536	_______	EA	Fixed Sash - Replace
__- ____	_______	__	____________________________________

DOORS

8- 3410	_______	EA	Rework Interior Door
8- 4376	_______	EA	Privacy Lock
8- 4392	_______	EA	Knob Set (Passage Lock)
8- 3326	_______	EA	Door Wrap & Lock
8- 3470	_______	EA	Door Stop
33- 3430	_______	EA	Door -- Flush Interior
8- 3402	_______	EA	Remove Interior Door
8- 4162	_______	EA	Door Trim Set 1x4
33- 3454	_______	EA	Wood Bifold
__- ____	_______	__	____________________________________

CLOSETS

8- 3474	_______	EA	Pole
8- 4154	_______	LF	Shelf and Pole
8- 4150	_______	LF	Wood Shelf
8- 4152	_______	EA	Standup Metal Closet
8- 4151	_______	EA	Closet Rack
8- 4532	_______	EA	Closet Bedroom

WALLS AND CEILINGS

15- 5134	_______	SF	Gut and Drywall - 1/2" W C
15- 5122	_______	SF	Shellcoat Drywall - 3/8th" W C
15- 5104	_______	SF	Drywall - Renail and Retape W C
15- 5109	_______	SF	Drywall - Patch Small Area W C
- 5003	_______	RM	Strip Paper, Spackle, Prime W C
15- 5004	_______	SF	Patch Plaster -- Large Area
15- 5164	_______	RM	Drywall Complete $_________________
__- ____	_______	__	____________________________________

PAINTING

15- 5203	_______	RM	Plaster Cracks $_________________
35- 5254	_______	EA	Prep & Paint Vacant Rm.
35- 5224	_______	SF	Prep & Paint Interior Trim

FLOORING

19- 5504	______	SF	Tile Flooring - Repair
8- 4210	______	SF	Floor Wood Patch
35- 5260	______	RM	Scrape and Paint Wood Floor
5- 5606	______	RM	Delete Carpet
19- 5602	______	SY	Carpet and Pad
19- 5519	______	RM	Underlay & VCT $__________
8- 4290	______	LF	Shoe Mold
8- 4286	______	LF	Baseboard - 1x4

STAIRS **FROM** __________ **TO**__________

ELECTRIC

23- 7084	__________	EA	3-Way Switch (pair)
23- 7108	__________	EA	3-Way Switch and Fixture
23- 7072	__________	EA	Replace Switch
23- 7094	__________	EA	Replace Light Fixture
23- 7126	__________	EA	Smoke Detector - Hard Wired
__- ____	__________	__	____________________

WALLS AND CEILINGS

15- 5134	__________	SF	Cut and Drywall - 1/2" W C
15- 5122	__________	SF	Shellcoat Drywall - 3/8th" W C
15- 5104	__________	SF	Drywall - Renail and Retape W C
15- 5108	__________	SF	Drywall - Patch
- 5003	__________	RM	Strip Paper, Spackle, Prime W C
15- 5005	__________	SF	Patch Plaster -- Small Area
15- 5010	__________	SF	Lime Skim Coat
_- ____	__________	__	____________________

STEPS AND HANDRAILS

8- 1520	__________	RI	Cellar Staircase Replace
8- 1550	__________	EA	Wood Handrail Repair
8- 1556	__________	EA	Handrail Brace
8- 1573	__________		Non-Skid Rubber Tread
8- 1508	__________	EA	Tread Repl. - Int
8- 1561	__________	LF	2 x 4 Inr Handrail
8- 1560	__________	LF	2" Round Handrail
8- 1562	__________	LF	Handrail with Balluster
8- 1562	__________	EA	Handrail with Ballusters
8- 1579	__________	EA	Staircase Repair - Custom $_____
_- ____	__________	_	____________________

FLOORING

19- 5504	__________	SF	Title Flooring - Repair
19- 5518	__________		Underlayment and VCT
8- 4210	__________	SF	Floor Wood Patch
8- 4220	__________	SF	Replace Damaged T & G
19- 5602	__________	SY	Carpet and Pad
5- 5606	__________	SF	Delete Carpet

WINDOWS

8- 3502	________	EA	Sash Lock
8- 3506	________	EA	Sash Cords
8- 3514	________	EA	Repair Window
8- 3574	________	EA	Glass Replace
10- 3518	________	EA	Weatherstrip Window
8- 3534	________	EA	New Window Sash
8- 3544	________	EA	Replace Douple Hung Window
__- ____	________	__	______________________

PAINTING

35- 5552	________	SF	Scrape & Paint Floor
35- 5259	________	EA	Paint Stairwell
__- ____	________	__	__________________________

ADDENDUM 6: MODEL CONTRACTOR CLASSIFICATION SYSTEM

Contractors are divided into 4 classes depending on size, licenses, past performance, and financial capacity.

1. <u>Class 1</u>

Limits:

a. Jobs up to $10,000.
b. Contractors may not carry more than $25,000 of outstanding work.

Requirements:

1. Application plus successful probation period.
2. Insurance
3. Continued timely completion of jobs.

2. <u>Class 2</u>

Limits:

a. Jobs up to $24,999.
b. Contractors may not carry more than $45,000 of outstanding work.

Requirements:

1. Application plus successful probation period.
2. Insurance
3. Continued timely completion of jobs.

3. <u>Class 3</u>

Limits:

a. Jobs to $50,000.
b. Contractor may not carry more than $75,000 of outstanding work.

Requirements:

All Class 1 and 2 plus -

1. State License.
2. Successful completion of five jobs or $75,000 of outstanding work.
3. Financial capacity to carry $12,000 week of work.

4. <u>Class 4</u>

Limits:

a. Jobs - unlimited.
b. Up to $200,000 of outstanding work.

Requirements:

All Class 1, 2, & 3 plus -

1. Financial capacity to carry $20,000 per week of work.

ADDENDUM 7a: CONTRACTOR RECRUITMENT TECHNIQUES

IDEAS	ACTION (Check as Appropriate) WILL TRY	IN PROGRESS
■ Inquire at building materials suppliers		
■ Ask surety bonding agents who sell bonds to subcontractors		
■ Inquire at commercial banks that make loans for subcontractors		
■ Talk with city building inspectors whose territory includes areas where private rehab work is ongoing		
■ Provide flyers to construction trade school and apprenticeship program instructors who teach subcontractors and/or prospective contractors		
■ Call local fire insurance agents		
■ Ask for a list of licensed subcontractors from the city bureau of licenses		
■ Call other city or private agencies who are engaged in rehab or housing management		

IDEAS	ACTION (Check as Appropriate) WILL TRY	IN PROGRESS
▪ Inquire at real estate agencies engaged in property management or rehabilitation		
▪ Visit private construction sites in areas of high residential rehab - copy phone numbers from trucks/talk with workers		
▪ Attend neighborhood meetings or civic associations - ask for contractor referrals		
▪ Place posters in building permits, licensing, and zoning offices		
▪ Contact and make presentations before local subcontractors' associations or any other organizations that include subcontractors		
▪ Accept loan client subcontractor referrals		
▪ Contact property management organizations		
▪ Consult yellow pages		
▪ Consult the Green Book of Home Improvement Contractors, available for 55 metropolitan areas		

ADDENDUM 7b: CONTRACTOR RECRUITMENT STATEMENTS

From the following statements, select those that your program can use to recruit contractors:

I. "We understand your need for quick payment..."

____ ▪ Unlike the private homeowner who may not pay, our program loans are placed in escrow at settlement. The money can be spent only on contract work

____ ▪ Our program normally pays for completed work within ____ working days

____ ▪ Our flexible draw schedules allow contractors to successfully manage their cash flow

____ ▪ The need for job financing (from a bank) is minimized through flexible progress payments

____ ▪ Other: ______________________________

II. "We can help you reduce your overhead costs..."

____ ▪ Bidding on our work reduces your advertising costs

____ ▪ We will protect you from costly and unfair homeowner disputes

____ ▪ Other: ______________________________

III. "We can keep your crews busy all year..."

____ ▪ Our jobs are not seasonal

____ ▪ Other: ______________________________

IV. "There is less red tape than you think..."

____ ▪ No payroll reports are required

____ ▪ Davis-Bacon does not apply for 7 units and under

____ ▪ Inspections are speedy and fair (We do not nit-pick!)

____ ▪ You sign a two-party (homeowner/contractor) contract. The city is not directly involved.

____ ▪ Our program is only a disbursing agent for the city government and HUD. City hall and HUD bureaucracies are not involved

____ ▪ Work can begin soon after settlement

____ ▪ Other: ______________________________

V. "We can provide you with technical assistance and training..."

____ ▪ Our program's technical library (including estimating manuals, construction material samples, etc.) is available for your use

____ ▪ Our job specifications take some of the uncertainty out of rehab work

____ ▪ Staff estimates can serve as a guide to you in making your bids

____ ▪ We can offer you training in the use and installation of new materials

____ ▪ We offer small business development training

____ ▪ Other: ______________________________

VI. "We are fair and impartial..."

____ ▪ Bidding procedures are designed to ensure fair competition

____ ▪ Our program will be an impartial referee in disputes

____ ▪ Other: ______________________________

VII. "We may be able to assist you with personnel problems..."

____ ▪ CETA workers are sometimes available

____ ▪ We can recommend quality subcontractors

____ Other: ______________________________

VIII. "We can help you with miscellaneous needs..."

____ ▪ Use our notary public for your legal documents without any charge

____ ▪ We will allow you to use office equipment (Xerox, mimeograph, typewriters) on a limited basis if you schedule the use in advance

____ ▪ Please have a cup of coffee on us

____ ▪ Other: ______________________________

ADDENDUM 7c: CONTRACTOR QUALIFICATION CHECKLIST

1. Contractor completed application form and filed with program

2. Contractor furnished certificate(s) of insurance

3. Program staff reviewed documentation

4. Program or lending institution staff verified information

   a. Creditor/business information

      - Dun and Bradstreet status reviewed
      - Suppliers interviewed
      - Subcontractors reviewed and interviewed

   b. Review of contractor's previous work

      - On-site visit(s) conducted
      - Telephone interview with past customer(s) completed

   c. Better Business Bureau contacted

   d. Consumer Protection Office contacted

5. Program staff reviewed contractor's file and decision was made for inclusion or rejection of contractor on Bidders' Probationary List

6. If awarded a contract, the Senior Rehab Staff reviewed first job

7. Program staff decides to accept or reject contractor for Permanent Bid List

ADDENDUM 8: REHABILITATION CONTRACTOR REGISTRATION APPLICATION

Please furnish the information requested below. This information will be used to determine your eligibility, and to verify information you supply for subcontractors and personnel.

BUSINESS NAME:______________________________

OWNER(S):______________________________

AUTHORIZED AGENT(S):______________________________

BUSINESS ADDRESS:______________________________PHONE NO.:______________________________

______________________________ ______________________________

DESCRIBE TRADES AND THE TYPES OF CONSTRUCTION IN WHICH YOUR COMPANY HAS EXPERIENCE.

____

____

____

LAST YEAR'S GROSS DOLLAR VALUE: $______________________________

HOW MANY JOBS HAVE YOU COMPLETED AS A GENERAL CONTRACTOR?: ______________________________

LARGEST DOLLAR AMOUNT: $______________________________

HOW LONG HAS YOUR COMPANY BEEN IN BUSINESS?: ______________________________

HAS THE COMPANY BEEN IN BUSINESS UNDER THE SAME NAME? YES______________ NO______________

IF THE COMPANY NAME HAS CHANGED, WHAT WAS THE ORIGINAL NAME(S)?

(COMPANY NAME) (ADDRESS) (DATE(S))

1)______________________________

2)______________________________

3)______________________________

HOW MANY EMPLOYEES DOES YOUR COMPANY HAVE? ______________________________

LICENSE NO.:______________________________

WHAT IS YOUR MAXIMUM CREDIT LINE AND WORKING CAPITAL AMOUNT?

MAXIMUM CREDIT LINE: $______________________________

WORKING CAPITAL: $______________________________

<u>REFERENCES</u>

WORK IN PROGRESS:

LIST THE NAMES, AND PHONE NUMBERS AND DOLLAR VALUE FOR THREE MOST RECENT JOBS COMPLETED OR IN PROGRESS.

NAME	PHONE NUMBER	DOLLAR VALUE
____________	( ) ____________	$ ________
____________	( ) ____________	$ ________
____________	( ) ____________	$ ________

PAST WORK:

LIST THE NAME, PHONE NUMBER AND DOLLAR VALUE AND COMPLETION DATE OF AT LEAST THREE CONTRACTS.

NAME	PHONE NO.	$ VALUE	COMPLETION DATE
____________	( ) ______	$ ______	____________
____________	( ) ______	$ ______	____________
____________	( ) ______	$ ______	____________

SUPPLIERS:

LIST TWO(2) MAJOR SUPPLIERS FROM WHOM YOU PURCHASE CONSTRUCTION MATERIALS. LIST NAME, AND PHONE NUMBER(S) AND CREDIT LIMIT OF EACH.

NAME	PHONE NUMBER	CREDIT LIMIT
____________	( ) ____________	$ ________
____________	( ) ____________	$ ________

LIST FINANCIAL INSTITUTIONS WITH WHOM YOU HAVE ESTABLISHED CREDIT, INCLUDE MAXIMUM CREDIT LINE.

	INSTITUTION	MAXIMUM CREDIT LINE
1.	______________________	$ ____________
2.	______________________	$ ____________
3.	______________________	$ ____________

PROVIDE A CURRENT LIST OF SUBCONTRACTORS WITH WHOM YOUR COMPANY HAS DONE BUSINESS. LIST NAME OF COMPANY, PHONE NUMBER(S), AND CONTACT PERSON FOR EACH SUB.

SUBCONTRACTOR	CONTACT PERSON	PHONE NO.
______________________	______________________	____________
______________________	______________________	____________
______________________	______________________	____________
______________________	______________________	____________
______________________	______________________	____________
______________________	______________________	____________

PLEASE SEND WITH THE APPLICATION A COPY OF THE FOLLOWING:

1. CURRENT LICENSE (FOR AS MANY TRADES AS YOU ARE CURRENTLY LICENSED)

2. CURRENT LIABILITY INSURANCE CERTIFICATE

3. CURRENT WORKMEN'S COMPENSATION CERTIFICATE

LIST ANY JOBS IN WHICH YOUR COMPANY HAS BEEN FOUND NEGLIGENT.

____

____

____

TO THE BEST OF MY ABILITY, I, THE UNDERSIGNED, ASSURE THAT THE INFORMATION LISTED ABOVE IS TRUE AND CORRECT. I FURTHER AUTHORIZE CHN TO VERIFY INFORMATION SUPPLIED AND OBTAIN A CREDIT REPORT.

______________________________ ______________________________

SIGNATURE OF OWNER DATE OF APPLICATION

SOCIAL SECURITY NO.____________________ OR FEDERAL I.D. NO ___________

CONTRACTOR'S ACCEPTANCE

The undersigned contracting firm agrees that in consideration for being placed upon the "Acceptable Contractor's Registry," he will comply with the following conditions on all rehabilitation work performed on properties located within the designated areas regardless of whether federal financing is or is not used by the owner:

1. To use only contract forms previously approved by the program.

2. That if work performed by the contractor is found to be unsatisfactory by this agency, or if contract relations between the contractor, homeowner or other parties are found to be unsatisfactory, that this agency may remove the contractor's name from the "Acceptable Contractor's Registry."

3. That work shall be performed in accordance with the Property Rehabilitation Standards, subject to such inspections as deemed necessary by this agency.

4. That minimum liability insurance will be provided.

5. Workmen's Compensation, where required, will be provided.

6. That the contractor will abide by Equal Opportunity provisions of the Civil Rights Act.

____
FIRM NAME

____
AUTHORIZED SIGNATURE

____
TITLE

____
DATE

ADDENDUM 9: REQUIREMENTS TO ENTER CONTRACTOR REGISTRY

1. Identification

Completed application form.

2. Insurance

Have your Insurance Agent send us your Certificate of Insurance showing you have Comprehensive Public Liability Insurance coverage protecting the property owner in the event of bodily injury, including death; and in the event of property damage arising out of work performed by the contractor, in the following amounts:

Type of Insurance	Each Occurrence	Aggregate
Bodily injury, including death	$100,000.00	$300,000.00
Property damage	$ 50,000.00	$100,000.00

AND

Evidence of Workman's Compensation Insurance.

NOTE: The current Certificate of Insurance should be mailed to the following address by the Insurance Company:

3. References:

Furnish a minimum of three (3) references from previous work contracted and completed. References must be from jobs that display your ability to successfully complete jobs like the jobs you are applying for through this office.

Example: If you are applying to bid on electrical work, give us references of wiring jobs you have completed.

A. Name(s) of person(s) dealt with.
B. Phone number(s) of same.
C. Job address(es) where work was performed.

4. Probationary Period:

Contractors meeting the above prerequisites shall remain on a probationary list until the contractor becomes the acceptable low bidder and satisfactorily completes one rehab job. Following this satisfactory completion, the contractor shall be placed on the Active Bidder's List as a Class 1 contractor and be eligible to bid on subsequent rehab jobs on a rotating, competitive basis. Should the contractor's performance during the probationary period be unsatisfactory, the contractor shall be removed from the Bidder's List and shall be ineligible to bid on future jobs.

ADDENDUM 10a: SUGGESTED QUESTIONS FOR INTERVIEWING CONTRACTORS

In the initial interview with the contractor, consider asking the following questions:

- How long have you been in business?
- In what areas and trades are you licensed by this jurisdiction? (Your local requirements will determine the evaluation of these answers.)
- How many employees do you employ full-time?
- Do you purchase material on credit or do you pay cash?
- If you had your choice of construction work, with which type do you feel most comfortable: New construction, remodeling, or rehab?

Determine the contractor's ability to perform by asking the following questions:

- What did you do prior to starting your construction business?
  - If you have no construction background, who in your company has it? Describe.
  - If your background is in a specialty (aluminum siding, windows, etc.), who has general knowledge of construction in your company? Describe his/her background.
- Do you work with your tools?
- How many workers do you employ and which trades do they cover? How long have they worked for you?
- Which trades are subcontracted?
  - Do you use the same subs on all jobs?
  - If not, do you get bids from several subs prior to submitting your quotation, or do you rely solely on your experience in estimating the job?
- For most of your work, do you write up the specifications and contract?
- Have you or has your company ever been in litigation?
- How do you respond to customer complaints?
  - During construction?
  - During close-out?
  - 6 months later?
  - 2 years later?

<u>Determine the contractor's ability to operate a business by asking the following questions</u>:

- Financial responsibility:
  - Do you maintain a separate bank account for business?
  - Do you have a running balance of working capital?
  - Have you established a line of credit with your bank to tide you over when collections are slow? To what limit will the bank go?
  - Who keeps the accounting records for your business?
  - Do you have other resources available to you? What are they?
  - Do you pay your withholding taxes on time?
  - Do you have a consulting attorney?
- What is the estimated value of your tools and equipment?
- How many trucks do you have?
- What other kinds of equipment do you have?
- Do you maintain an office or do you work out of your home?
- Who answers your business phone? (Check one)
  - Full-time employee?
  - Part-time employee?
  - Relative?
  - Electronic answering device?
- If you're working in the field, how many times a day do you call in for messages?

ADDENDUM 10b: SUGGESTED QUESTIONAIRE FOR INTERVIEWING CONTRACTOR'S PAST CUSTOMERS

CUSTOMER'S NAME:____

1. Were you satisfied with work performed? Yes No

2. Was the job completed on time? Yes No

3. Were the workers competent and agreeable? Yes No

4. Did they clean up at the end of each day? Yes No

5. Were you satisfied with the work performed by the contractor(s)? Yes No

6. If problems arose, was the contractor receptive to discussion? Yes No

7. What type of work did the contractor perform?________________

____

____

8. What was the scope of the job?______________________________

____

____

9. On a scale of 1 to 10, how would you rate his overall performance on the job? (Circle)

1 2 3 4 5 6 7 8 9 10

10. Would you hire the contractor again? Yes No

ADDENDUM 10c: QUESTIONNAIRE--
CONTRACTOR'S BANKING REFERENCES

1. How long has contractor or individual been doing business with your bank?

2. What kind of accounts associated with construction financing, development, or mortgages has contractor or individual been involved with at your bank?

3. How would you rate the contractor's or individual's repayment history relative to such bank loans?

4. What kind of credit line does the bank afford or make available to such construction firms or individuals?

5. If you were asked to do so, would you have enough knowledge of this firm or individual to recommend them for a construction project in the range of $500,000 to $1,000,000?

ADDENDUM 10d: QUESTIONNAIRE FOR CONTRACTOR'S SUPPLIER (PRODUCTS) REFERENCES

1. What is the range of building and/or construction products, materials that you supply to the building trades?

2. How long has the contractor and/or individual been doing business with you?

3. What kind of credit line do you extend to your contractor customers?

   A. What is the ceiling maximum for credit?

   B. What kind of credit terms do you allow?

      1) 30 days
      2) 45 days
      3) 60 days
      4) 90 days
      5) More

   C. What do you do concerning outstanding balances?

4. What kind of credit history does the contractor and/or individual have with you?

5. How is the contractor's and/or individual's repayment track record?

ADDENDUM 10e: VERIFICATIONS OF CONTRACTOR'S LISTED PROJECTS

Step #1. Verify contractor's corporation, individuals, or other information with legal department or Office of Consumer Affairs.

Step #2. Check contractor's track record with construction officials, and building inspectors in the municipalities in which referenced projects are located.

Step #3. Field check referenced projects that are:

A. In progress:

Project Description	Location
1) ____________________	____________________
2) ____________________	____________________
3) ____________________	____________________

B. Completed within the past three to six months:

Project Description	Location
1) ____________________	____________________
2) ____________________	____________________
3) ____________________	____________________

ADDENDUM 10f: SUGGESTED FORM FOR THE FIELD INSPECTION OF CONTRACTOR'S WORK

Project: ______________________________ Owner: ______________________________

______________________________ ______________________________

Role Of Contractor: ____

____

TIME: ___ On schedule ____ Ahead of schedule Why? ______________________

____ Behind schedule ______________________

____

MONEY: ___ On budget ____ Over budget Why? ______________________

____ Under budget ______________________

QUALITY: Materials - to specifications? (Circle) Yes No

Workmanship: (Check one)

Standard for job_____ Substandard_____ Above standards_____

ADMINISTRATION: (To Date) ____

____

____

Would you reuse this contractor? (Circle) Yes No

WHY? ____

____

____

ADDENDUM 11: PRODUCTION DEPARTMENT
JOB FILE INDEX

BORROWER(S):____

ADDRESS: ____

TELEPHONE: ____________ HOME:___________ WORK:(Borrower)________________

____________ ___________ (Co-Borrower)________________

ACTIVITY DESCRIPTION		DATE REQUESTED	DATE RECEIVED	COMMENTS
Pre-Application				
Verification of Employment	(a)			
Verification of Employment	(b)			
Verification of Mortgage				
Verification of Loans(s)				
Verification of Income				
Credit Report				
Title Opinion				
Work Write-Up				
Appraisal				
Owner's Approval of Work Write-Up				
Bid Proposal				
Bid Review				
Contractor Selection by Borrower				
Completed Loan Application				
Approval by Field Office				
Loan Set-Up Form				
Transmittal to Lender (Initial Settlement)				
Executed Rehabilitation Contract				
Proceed Order				
Request for Progress				
Payment Drawdown	1.)			
	2.)			
	3.)			
	4.)			
Request for Final Drawdown				
Final Waiver of Liens				
Certificate of Final Inspection				
Closeout Documentation forwarded to Lender				

FINAL COMMENTS:____

____

____

____

ADDENDUM 12: WORK-IN-PROGRESS

FILE INDEX-SFSR

DATE	DOCUMENT	INITIAL
________		________
________	Job Log	________
________	Violation Notice (if available)	________
________	Construction Document Checklist	________
________	Building Inspection	________
	Specification Writing and Estimating	
________	Work Write-up and Estimating	________
________		________
	Bidding	
________	Invitation to Bid	________
________	Instructions to Bidders	________
________	Bid Form	________
________	Bid Award	________
________	Bid Results	________
	Contracts	
________	Job Cost Breakdown	________
________	Sworn Statement	________
________	Homeowner/Contractor Agreement	________
________	Homeowner/___________ Agreement	________
________	General Conditions	________
________	Loan/Construction Escrow Agreement	________
________	Construction Loan Contingency Escrow Agreement	________
________	Building Permit	________
________	Notice to Proceed	________
	Construction Monitoring	
	Payment #1	
________	Partial Inspection and Payout Order	________
________	Sworn Statement	________
	Waiver of Lien and Contractor's Affidavit	
________	General	________
________	Subcontractor	________
________	Payment Release	________
________	Field Order	________
________	Change Order	________

ADDENDUM 12

DATE	DOCUMENT	INITIAL
	Payment #2	
________	Partial Inspection and Payout Order	________
________	Sworn Statement	________
	Waiver of Lien and Contractor's Affidavit	
________	General	________
________	Subcontractors	________
________	Payment Release	________
________	Field Order	________
________	Change Order	________
	Payment #3	
________	Partial Inspection and Payout Order	________
________	Sworn Statement	________
	Waiver of Lien and Contractor''s Affidavit	
________	General	________
________	Subcontractors	________
________	Payment Release	________
________	Field Order	________
________	Change Order	________
	Project Closeout	
________	City Final Inspection	________
________	Punch List	________
	Final Payment #4	
________	Field Order	________
________	Change Order	________
________	Final Inspection and Payout Order	________
________	Sworn Statement	________
	Waiver of Lien and Contractor's Affidavit	
________	General	________
________	Subcontractors	________
________	Payment Release	________
________	Certificate of Completion	________
	Construction Follow-up	
________	Contractor Evaluation	________

ADDENDUM 13: SAMPLE SPECMASTER REPORTS

WORK WRITE UP pp.1-2

THE WORK WRITE UP INCLUDES:

- TRADE & SPEC NUMBER
- SLUG
- FULL SPEC
- ADDENDA
- QUANTITY
- UNIT

AVAILABLE WITH OR WITHOUT PRICES

SPECIFICATIONS BY LOCATION pp.3-4

SPECS BY LOCATION INCLUDES:

- SPECS & TRADE NUMBER
- SLUG
- ADDENDA
- QUANTITY
- UNIT

AVAILABLE WITH OR WITHOUT PRICES

COST ESTIMATE p.5

THE COST ESTIMATE INCLUDES:

- SPEC & TRADE NUMBER
- SLUG
- TOTAL QUANTITY PER LINE ITEM
- TOTAL COST PER LINE ITEM
- SUBTOTALS BY TRADE
- GRAND TOTAL AT END OF REPORT

DETAILED SPECIFICATIONS BY TRADE p.6-8

DETAILED SPECS BY TRADE INCLUDES:

- TRADE & SPEC NUMBER
- SLUG
- FULL SPEC
- LOCATIONS AND ADDENDA
- QUANTITY FOR EACH LOCATION
- UNIT

AVAILABLE WITH PRICES OR WITH BLANKS FOR BIDDING

ADDENDUM 13

Work Write-Up

Date: 08-25-89 11:02 AM
Project: CDBG Address: 1111 East Main St. Unit: 1

Item No.	Item	Quan.	Unit	Cost
General Requirements				
1-4	PERMITS AND FEES Apply for, post on job site and pay all required fees for building permit.	1	EA	$ 36
1-54	JOB SITE SIGN Post owner-supplied sign on front of building. Return to owner at final payment.	1	EA	$ 20
1-70	DEBRIS REMOVAL Dispose of all debris from worksite and rear yard in code legal dump.	2	EA	$200
25-7626	LEAD PAINT HEPA VACUUM Thoroughly vacuum ALL VISIBLE SURFACES in structure with a vacuum equipped with a High Efficiency Particle Accumulator (HEPA), including front and rear porches and steps and accessible duct work.	1	DU	$200
Front Exterior				
6-882	SIDEWALK REPAIR Remove portions of deteriorating sidewalk down to solid surface. 4" minimum depth. Resurface with a patching cement mixture. Bond with proper bonding agent. Dispose of rubble properly.	36	SF	$126
10-2776	STORM WINDOW/SCREEN--CAULKED Install a white aluminum combination storm and screen, self-storing, triple track, fully weatherstripped and caulked.	4	EA	$240
14-3342	DOOR--EXTERIOR PANELED-FRONT Install a 1-5/8", 6-panel wood door with 3 butt hinges, mail slot, peep hole, vinyl sweep, vinyl weatherstripping, and owner-supplied entry lock. Re-use existing dead bolt. Paint door.	1	EA	$250
14-4340	HOUSE NUMBERS Install 3" high metal or PVC house numbers on a 1"x 4" pine board painted white with 2 coats of exterior paint.	1	EA	$ 26
24-7504	RAILING--WROUGHT IRON--STEPS Design, fabricate and install an economy steel railing of 2" flat top and bottom rail and 1/4" ballustrades 6" on center.	1	EA	$150

Work Write-Up

Date: 08-25-89 11:02 AM
Project: CDBG Address: 1111 East Main St. Unit: 1

Item No.	Item	Quan.	Unit	Cost
1st Floor Hallway				
14-3410	DOOR--REWORK INTERIOR Rehang and plane door, and adjust and door and lockset to operate properly. INNER FRONT DOOR.	1	EA	$42
15-5004	PATCH PLASTER Cut back damaged plaster. Cut out cracks 1/4" wide in a vee joint. Renail all loose lath. Install 1/8" flat rib metal lath where wood is not reusable. Apply basecoat, allowing at least 1/16" for finish coat. After 24 hour cure, apply finish coat.	40	SF	$60
17-5254	DECORATE VACANT ROOM Cover hardware/fixtures not to be painted. Scrape loose/peeling surfaces, incl. floor, trim closet. Feather edges. Clean grease, fungus, dust from surfaces. Fill holes/cracks. Paint to cover: Trim/kitchen/bath w/ oil enamel; floor w porch & deck enamel; all other surfaces w/ latex. Contractor may substitute vinyl paper or water-resistant paneling in bath & kit, & residential paper or paneling elsewhere. Color choice by owner.	1	RM	$95
19-5518	UNDERLAY AND VCT Repair and/or level floor as necessary. Install 1/4" underlayment grade plywood using 7d screw shank or cement coated nails, 6" on center in all directions. Lay 12"x 12"x 3/32" vinyl composition tile, color group B as made by Armstrong or Azrock, per manufacturer's recommendations. Include metal edge strips at openings, and shoe or vinyl base around perimeter. Owner's choice of in-stock color. IN VESTIBULE NEW SUBFLOOR.	8	SF	$16
Living Room				
14-3410	DOOR--REWORK INTERIOR Rehang and plane door, and adjust door and lockset to operate properly.	1	EA	$42
14-3470	DOOR STOP Install baseboard mounted solid metal door stop.	1	EA	$ 5

General Specifications by Location

Date: 08-25-89   11:04 AM
Project: CDBG   Address: 1111 East Main St. Unit: 1

Item No.	Item	Quan.	Unit	Cost
General Requirements				
1-4	PERMITS AND FEES	1	EA	$ 36
1-54	JOB SITE SIGN	1	EA	$ 20
1-70	DEBRIS REMOVAL	2	EA	$200
25-7626	LEAD PAINT HEPA VACUUM	1	DU	$200
Front Exterior				
6-882	SIDEWALK REPAIR	36	SF	$126
10-2776	STORM WINDOW/SCREEN--CAULKED	4	EA	$240
14-3342	DOOR--EXTERIOR PANELED-FRONT	1	EA	$250
14-4340	HOUSE NUMBERS	1	EA	$ 26
24-7504	RAILING--WROUGHT IRON--STEPS	1	EA	$150
1st Floor Hallway				
14-3410	DOOR--REWORK INTERIOR INNER FRONT DOOR.	1	EA	$ 42
15-5004	PATCH PLASTER	40	SF	$ 60
17-5254	DECORATE VACANT ROOM	1	RM	$ 95
19-5518	UNDERLAY AND VCT IN VESTIBULE NEW SUBFLOOR.	8	SF	$ 16
Living Room				
14-3410	DOOR--REWORK INTERIOR	1	EA	$ 42
14-3470	DOOR STOP	1	EA	$ 5
15-5004	PATCH PLASTER	6	SF	$ 35
17-5254	DECORATE VACANT ROOM	1	RM	$ 95
23-7072	SWITCH REPLACE IN HALL.	2	EA	$ 24
Dining Room				
14-3506	SASH CHAINS	1	PR	$ 17
14-3574	GLASS REPLACE--WOOD SASH	1	EA	$ 17
16-5104	DRYWALL--RENAIL & RETAPE CEILING.	130	SF	$ 97
17-5254	DECORATE VACANT ROOM	1	RM	$ 95

ADDENDUM 13
Page 4

General Specifications by Location

Date: 08-25-89 11:04 AM
Project: CDBG Address: 1111 East Main St. Unit: 1

Item No.	Item	Quan.	Unit	Cost
Dining Room				
21-6170	GAS SPACE HEATER--REMOVE	1	EA	$ 25
23-7004	RECEPTACLE REPLACE	1	EA	$ 12
23-7010	RECEPTACLE-15 AMP FISHED	1	EA	$ 50
23-7094	LIGHT FIXTURE--REPLACE	1	EA	$ 38
Kitchen				
14-3410	DOOR--REWORK INTERIOR TO BASEMENT.	1	EA	$ 42
14-3470	DOOR STOP	1	EA	$ 5
14-3506	SASH CHAINS AT REAR.	2	PR	$ 34
14-3510	WINDOW SASH--REPAIR	2	EA	$ 32
14-3574	GLASS REPLACE--WOOD SASH	4	EA	$ 68
16-5162	DRYWALL--FUR, HANG, FINISH AT CEILING.	160	SF	$272
17-5254	DECORATE VACANT ROOM	1	RM	$ 95
18-5420	PANELING--WATERPROOF TO 4' HIGH.	128	SF	$205
19-5518	UNDERLAY AND VCT	160	SF	$328
22-6232	SUPPLY--COPPER	20	LF	$108
22-6374	GAS SHUT-OFF VALVE	1	EA	$ 20
22-6394	CAB--METAL BASE, SINK, R. DRAIN IN CORNER.	1	EA	$525
23-7014	RECEPTACLE--WIRE 20 AMP	2	EA	$104
23-7016	RECEPTACLE-20 AMP COUNTERTOP	1	EA	$ 52
23-7092	LIGHT FIXTURE AND SWITCH	1	EA	$ 95
Basement				
14-3570	REGLAZE SASH	2	EA	$ 16
17-5262	PREP AND PAINT STAIRWELL WALLS ONLY.	1	EA	$ 60
21-5828	FURNACE/DUCT--GAS: 2 STORY, 2 BR	1	EA	$1,600

ADDENDUM 13
Page 5

Cost Estimate

Date: 08-25-89   11:05 AM
Project: CDBG   Address: 1111 East Main St. Unit: 1

Item No.	Item	Quan.	Unit	Labor	Mater.	Subs	Total
1 General Requirements							
1-4	PERMITS AND FEES	1	EA	0	0	36	$ 36
1-54	JOB SITE SIGN	1	EA	0	0	20	$ 20
1-70	DEBRIS REMOVAL	3	EA	0	0	300	$300
	Subtotals			0	0	356	$356
4 Site Work							
4-310	FENCE--CHAIN LINK 4' HIGH	33	LF	0	0	198	$198
4-360	GATE--GALVANIZED 4' HIGH	1	EA	0	0	65	$ 65
4-540	T-POLE	2	EA	0	0	80	$ 80
	Subtotals			0	0	343	$343
6 Concrete and Paving							
6-876	CONCRETE SLAB--4"	20	SF	0	0	70	$ 70
6-882	SIDEWALK REPAIR	36	SF	0	0	126	$126
	Subtotals			0	0	196	$196
8 Rough Carpentry							
8-1556	HANDRAIL BRACE	2	EA	0	0	30	$ 30
	Subtotals			0	0	30	$ 30
9 Roofing							
9-2304	ROOF--DOUBLE COVERAGE SIS	7.5	SQ	0	0	600	$600
	Subtotals			0	0	600	$600
10 Conservation							
10-2776	STORM WINDOW/SCREEN--CAULKED	9	EA	0	0	540	$540
	Subtotals			0	0	540	$540
14 Finish Carpentry							
14-3340	DOOR--EXTERIOR FLUSH-REAR	1	EA	0	0	150	$150
14-3342	DOOR--EXTERIOR PANELED-FRONT	1	EA	0	0	250	$250
14-3410	DOOR--REWORK INTERIOR	7	EA	0	0	294	$294
14-3470	DOOR STOP	5	EA	0	0	25	$ 25
14-3502	SASH LOCK	4	EA	0	0	20	$ 20
14-3506	SASH CHAINS	7	PR	0	0	119	$119

Detailed Specifications by Trade

Date: 08-25-89   10:58 AM
Project: CDBG   Address: 1111 East Main St. Unit: 1

Item No.	Item	Quan.	Units	Unit Price	Total Price
	14 Finish Carpentry				
3340	DOOR--EXTERIOR FLUSH-REAR				
	Install solid core, flush panel, exterior door with keyed lock set, three 3"x4" butt hinges, sealing threshold, & bronze weatherstripping.				
	Rear Exterior				
3342	DOOR--EXTERIOR PANELED-FRONT	1	EA	$150	$150
	Install a 1-5/8", 6-panel wood door with 3 butt hinges, mail slot, peep hole, vinyl sweep, vinyl weatherstripping, and owner supplied entry lock. Re-use existing dead bolt. Paint door.				
	Front Exterior	1	EA	$250	$250
3410	DOOR--REWORK INTERIOR				
	Rehang and plane door, and adjust door and lockset to operate properly.				
	Living Room	1	EA	$ 42	$ 42
	Rear Bedroom	1	EA	$ 42	$ 42
	Front Bedroom	2	EA	$ 42	$ 84
	Kitchen				
	TO BASEMENT	1	EA	$ 42	$ 42
	1st Floor Hallway				
	INNER FRONT DOOR.	1	EA	$ 42	$ 42
	Bathroom	1	EA	$ 42	$ 42
3470	DOOR STOP				
	Install baseboard mounted solid metal door stop.				
	Living Room	1	EA	$ 5	$ 5
	Front Bedroom	2	EA	$ 5	$ 10
	Bathroom	1	EA	$ 5	$ 5
	Kitchen	1	EA	$ 5	$ 5
3502	SASH LOCK				
	Install brass plated sash lock on double hung window.				
	Rear Bedroom	1	EA	$ 5	$ 5
	Bathroom	1	EA	$ 5	$ 5
	Front Bedroom	2	EA	$ 5	$ 10

Item No.	Item	Quan.	Units	Unit Price	Total Price
3506	SASH CHAINS				
	Install sash chains between sash and counterweights. Adjust window to operate smoothly.				
	Dining Room	1	PR	$17	$17
	Kitchen	2	PR	$17	$34
	Bathroom	1	PR	$17	$17
	Rear Bedroom	1	PR	$17	$17
	Front Bedroom	2	PR	$17	$34

Bidder:______________________________ Total: ______________

Detailed Specifications by Trade

Date: 08-25-89 10:58 AM
Project: CDBG Address: 1111 East Main St. Unit: 1

Item No.	Item	Quan.	Units	Unit Price	Total Price
16	Drywall				
5104	DRYWALL--RENAIL & RETAPE Renail or screw gypsum surface 6" on center. Scrape seams and nail pops. Retape where tape is missing, damaged or bubbled. Finish with 2 coats of compound.				
	Dining Room	130	SF	______	______
5124	LAMINATE 1/2" DRYWALL Install 1/2" gypsum drywall over existing plaster surface with screws 8" on center and a bead of construction adhesive 20" on center. On walls remove top molding from 3 piece base and reinstall after surface is paint-ready. Tape and 3 coat finish.				
	Rear Bedroom AT WALL BETWEEN MIDDLE AND FRONT BEDROOMS AND BATH DOOR WALL, AND CEILING.	380	SF	______	______
5162	DRYWALL--FUR, HANG, FINISH Fur surface with 1"x3" wood or metal furring strips. Hang, tape and 3 coat finish 1/2" drywall, glued and nailed or screwed 8" on center.				
	Kitchen	160	SF	______	______

Bidder:________________________________ Total: ______________

ADDENDUM 14: INVITATION TO BID

DATE: ____________________

JOB ADDRESS: ____

PERSON TO CONTACT TO GAIN ENTRY:

NAME: ____

ORGANIZATION: ____

PHONE: ____

BIDS DUE: ______________________________________

SUBMIT BIDS TO: ________________________________

____

Dear Contractor,

You are invited to submit a proposal for rehabilitation as specified in attached Work Write-Up Report. The property must be carefully examined and all quantities and field conditions verified. Please submit the "Bid Proposal Form" showing the following:

- Bid price for entire job.
- Earliest starting date.
- Proposed completion date.
- Any addenda showing work not covered on our Work-Write Up Report.

The program reserves the right to reject any oral bids and require any bidder to unit price the proposal.

ADDENDUM 15: BID PROPOSAL FORM

TO: ______________________________ DATE SUBMITTED: __________________

PROGRAM MANAGER

PROPERTY ADDRESS: ________________

____

The following proposal is hereby submitted covering all Work shown and/or specified in the attached bid documents received for this job, which are:

1. WORK Write-Up __________ Dated ____________________
2. ________________________________ Dated ____________________
3. ________________________________ Dated ____________________
4. ________________________________ Dated ____________________

The undersigned, in having become thoroughly familiar with the terms and conditions of the proposed Contract Documents and with local conditions affecting the execution of the Work, proposes to complete the Work for the sum of ________________dollars $ ______________________.

Within the following time frame:

Start within _____________ days of notice to proceed and complete within _____________ calendar days thereafter.

Upon Request by the AGENCY, contractor agrees to itemize any/or all aspects of this proposal.

Contractor: ______________________________ ______________________________ (Signature)

Address: ______________________________

______________________________ ______________________________ (Date)

Phone: ______________________________

Employer ID # ______________________________

Proposed Addenda:

The Contractor further proposed the following changes, deletions, or additions to the Write-up.

PRICE

____ $____________________

____ $____________________

____ $____________________

ADDENDUM 16: OWNER SELF-HELP
PERFORMANCE AGREEMENT

THIS AGREEMENT made and entered into as of the ________ day of ______________, 19_____, by and between ________________________________(the "Owner/General contractor"); and ________________________________(the "Supervisor"). (The Housing Program.)

WITNESSETH:

WHEREAS, the Housing and Community Development Act of 1974, as amended, provides funds (the "Funds") for rehabilitation in Federally Assisted Urban Renewal and Community Development Block Grant areas;

WHEREAS, Owner/General Contractor desires to use the benefits of said Act for the purpose of performing the work hereinafter described; and

WHEREAS, Supervisor desires that such work be done and agrees to make certain the work is done in a workmanlike manner, within budget and on time;

NOW, THEREFORE, in consideration of the agreements hereinafter set forth, the receipt and sufficiency of which are hereby acknowledged, the parties agree as follows:

ARTICLE I

DESCRIPTION AND SCOPE OF WORK

Owner/General Contractor shall perform or cause to be performed the work (the "Work") shown on the drawings (the "Drawings"), if any, and described in the specifications dated ___________________________, items ________________ through __________________, attached hereto as EXHIBIT "A" on the property at _______________________ REPORT (EXHIBIT "B") attached hereto. Owner/General Contractor shall furnish or cause to be furnished all of the labor, materials, tools, equipment, construction equipment, and machinery to be used in the performance of the Work.

ARTICLE II

TIME AND COMMENCEMENT AND COMPLETION

The Work shall be commenced within _________________________ days after ________________________, 19______, which date will constitute the Proceed Date. If the Work is not commenced within period specified, Supervisor, at Supervisor's election, may terminate this Agreement by written notice to Owner/General Contractor. After commencement, Owner/General Contractor shall complete the Work within ___________ days of the Proceed Date. (NOTE: Construction time not to exceed 180 calendar days.)

ARTICLE III

TOTAL PRICE

Owner/General Contractor has obtained subcontractor bids on certain elements of the Work and estimates of material costs to be used by the Owner/General Contractor for the balance of the Work, and these total ______________________________ ($______________) for subcontractors and ______________________________ ($______) for materials, a total overall cost of______________________________ ______________________________($______________).

ARTICLE IV

PROGRESS PAYMENTS

The Funds being made available for the Work will be held in escrow by Supervisor, and will be disbursed by Supervisor in accordance with this Agreement. Supervisor shall make payments on account of the total price to Owner/General contractor, as requisitions are submitted and in accordance with the construction agreement, approved by Supervisor, but in no instance more often than once a month.

If the Work has been substantially completed but full completion is delayed through no fault of Owner/General Contractor, Owner/General Contractor shall receive payment for no more than ninety percent (90%) of the value of the completed, inspected, and accepted Work as determined by Supervisor, and the balance of the Funds for the Work shall remain in escrow for release upon full completion, inspection, and acceptance of the Work by Supervisor. If Owner/General Contractor is unable to complete the Work in the time frame specified in ARTICLE II (as may be extended in accordance with this agreement) and after written notice, is unable to show within 30 days substantial and satisfactory progress, Supervisor may at his sole discretion select an approved contractor to complete the Work. As the cost for completion of the Work by the selected contractor may exceed the balance of funds available, Owner/General Contractor shall be assessed this difference.

ARTICLE V

ACCEPTANCE AND FINAL PAYMENT

Final payment of the total price shall be due after completion of the Work, provided Owner/General Contractor has delivered to Supervisor a complete release of all liens arising out of this Agreement or receipts evidencing payment in full of all labor, materials and equipment for which a lien could be filed, or in the alternative a bond satisfactory to Supervisor indemnifying all parties against such liens.

ARTICLE VI

CHANGES IN THE WORK

After this Agreement is executed by Supervisor, any changes to the Specifications, if any, and the Drawings, if any, must be agreed upon in writing by Supervisor.

ARTICLE VIII

GENERAL CONDITIONS

Owner/General Contractor. Owner/General Contractor shall supervise and direct the Work, using Owner/General Contractor's best skill and attention, and Owner/General Contractor shall be solely responsible for all construction, means, methods, techniques, sequences and procedures and for coordination of all portions of the Work.

Governing Law. This Agreement shall be governed in accordance with the laws of the State of ____________________________.

Successors and Assigns. The provisions of this Agreement shall inure to the benefit and be binding upon the parties hereto, and their respective heirs, representatives, successors and assigns.

TIME. All time limits stated in this Agreement are of the essence. If Owner/General Contractor is delayed at any time in the progress of the Work by changes ordered in the Work, by labor disputes, fire, unusual delay in transportation, unavoidable casualties, causes beyond Owner/General Contractor's control, or by any other cause which justifies the delay, then the time set for completion of this Contract, pursuant to ARTICLE II, shall be extended for periods of not more than thirty (30) days as agreed upon in writing by Owner/General contractor and Supervisor.

Warranty. Owner/General Contractor warrants that for a period of one (1) year following completion of the Work, the Work will be of good quality, free from faults and defects.

Taxes, Permits, Fees and Notices. Owner/General Contractor shall pay all sales, consumer, use and other similar taxes which are legally in force during the time the work is carried out, and secure and pay for all permits, governmental fees, licenses and inspections necessary for the proper execution and completion of the Work. Owner/General Contractor shall give all notices and comply with all laws, ordinances, rules, regulations and lawful orders of any public authority bearing on the performance of the Work.

Owner/General Contractor's Liability Insurance. Owner/General Contractor shall purchase and maintain such insurance as will protect him from claims under workmen's compensation acts and other employee benefit acts, from claims for damages because of bodily injury, including death, and from claims for damages to property, other than to the Work, which may arise out of or result from the Owner/General Contractor's operations under this Agreement, whether such operations be by himself or by any subcontractor or anyone directly or indirectly employed by any of them.

Property Insurance. Owner/General Contractor shall purchase and maintain property insurance upon the work at the Property to the full insurable value thereof. This insurance shall protect the interests of Owner/General Contractor, subcontractors and sub-subcontractors relating to the Work and shall insure against perils of fire and extended coverage and shall include "all risk" duplication of coverage, vandalism, theft, and malicious mischief. Any insured loss is to be adjusted with Owner and made payable to the Owner as trustee for the insured, as their interests may appear, subject to the requirements of any mortgage clause.

Executive Order 11246. Executive Order 11246 shall apply and is made a part of this Agreement if the sum to be charged for the Work is Ten Thousand Dollars ($10,000.00) or more.

Lead-Based Paint. Owner/General Contractor shall, during the performance of the Work, eliminate any existing lead-based paint hazards on applicable surfaces in the Work, in accordance with the Lead-Based Paint Poisoning Prevention Act, as amended (42 U.S.C. Section 4801 et seq.) and the Department of Housing and Urban Development Lead-Based Paint Regulations, as amended (24 C.F.R. Part 35), promulgated pursuant hereto.

Inspection. Owner/General Contractor shall permit Supervisor, the Local Housing Authority, the U.S. Government or any of their designees or representatives to examine and inspect the Work.

Subcontractors. Owner/General Contractor shall be responsible for the acts and omissions of subcontractors and their agents and employees, and other persons performing any of the Work under an agreement with Owner/General Contractor.

Conflict of Interest. No member of or delegate to the Congress of the United States and no Resident Commissioner shall be admitted to any share or part of this Agreement, proceeds of the loan being made to Owner for the Work, or to any benefit to arise from the same. No officer, employee, or member of Supervisor which exercises any functions or responsibilities in connection with this carrying out of the project to which this Agreement pertains shall have any private interest, direct or indirect, in this Agreement.

<u>Indemnification of Supervisor</u>. Owner/General Contractor agrees to indemnify and forever hold and save harmless Supervisor from any and all losses, costs, injuries or damages of every kind, nature or description arising under this Agreement or from any source, whatsoever, except any losses, costs, injuries, or damages resulting from the sole negligence of Supervisor or its agents or employees.

<u>Whole Agreement</u>. The parties acknowledge and agree that this Agreement constitutes the entire agreement between the parties. This Agreement may be amended or modified only by a written instrument signed by or on behalf of the parties hereto.

IN WITNESS WHEREOF, the parties have executed this Agreement as of the date and year first above written.

____________________	____________________
(OWNER/GENERAL CONTRACTOR)	(SUPERVISOR)
____________________	____________________
(WITNESS)	(WITNESS)
____________________	____________________
(DATE)	(DATE)

ADDENDUM 17: AGREEMENT BETWEEN OWNER AND CONTRACTOR

Recommended Contract Modifications
2/13/89

AGREEMENT BETWEEN OWNER & CONTRACTOR

AGREEMENT: made as of the ______________________
Date

BETWEEN the Owner

and the Contractor

Name: ______________________

Address: ______________________

____

Project Address: ______________________

____

CONTRACT DOCUMENTS

Contract Documents, which constitute the entire agreement between the Owner and the Contractor and are as fully a part of the Contract as if attached, are enumerated as follows:

(Delete any that are not applicable to this project).

1. This "Agreement and General Conditions."
2. "Procedures for Contractors."
3. Detailed Work Description and Itemized Bid Dated ______________ ("Specifications and Bid").
4. General Specification Manual.
5. Addenda No. __________
6. Attached sketches/drawings.
7. Owner selection list Dated ________________
8. Other______________________________

THE WORK

The contractor shall perform the entire rehabilitation of the residential structure as described in the contract documents except as indicated as follows to be the responsibility of others:

Scope	Contractor

TIME OF COMMENCEMENT & SUBSTANTIAL COMPLETION:

The Work shall commence within 7 calendar days of authorization by written Notice to Proceed from the Owner.

The Work shall be substantially completed no later than ___ calendar days from the date of the Notice to Proceed. The Contractor shall be liable for and shall pay the owner ____$_________ as liquidated damages for each calendar day of delay until the work is substantially completed.

Optional

[If Work is delayed at any time by causes beyond the Contractor's control, then the Contract may be extended for such reasonable time as the Owner's Authorized Representatives may determine.]

OWNER'S REPRESENTATIVES

The Owner's Representatives shall be

____ (Construction Manager)

____ (Neighborhood Director)

The Owner's Representatives will provide administration of this Contract during construction and throughout the warranty period.

The Owner's Representatives will visit the site at intervals appropriate to the stage of construction to determine if the Work is proceeding in accordance with the Contract Documents.

Based on Owner's Representative's evaluation of Contractor's invoices for payment, the Owner's Representative will determine the amounts owing to the Contractor.

The Owner's Representatives will have authority to reject Work that does not conform to the Contract Documents.

If the Contractor fails to correct defective Work or persistently fails to carry out the Work in accordance with the Contract Documents, the Owner's representatives, by a written order, may order the Contractor to stop the Work, or any portion thereof, until the cause for such order has been eliminated.

CONTRACTOR RESPONSIBILITIES

The Contractor shall supervise and direct the Work, using his best skill and attention, and he shall be solely responsible for all construction means, methods, techniques, sequences and procedures and for coordinating all portions of the Work under the Contract.

The Contractor warrants to the Owner that all materials and equipment incorporated in the Work will be new unless otherwise specified, and that all Work will be of good quality, free from faults and defects and in conformance with the contract Documents. All Work not conforming to these requirements may be considered defective.

The Contractor shall give all notices and comply with all laws, ordinances, rules, regulations, and lawful orders of any public authority bearing on the performance of the Work, and shall promptly notify the Owner's Representatives if the Drawings and Specifications are at variance therewith.

The Contractor shall be responsible for all safety precautions in connection with this Work. He shall take all legally required and reasonable precautions for the safety of all employees on the Work and other persons who may be affected thereby.

To the fullest extent permitted by law, the Contractor shall indemnify and hold harmless the Owner and the Owner's Representatives and their agents and employees from and against all claims, damages, losses and expense, including but not limited to attorneys' fees arising out of or resulting from the performance of the Work, provided that any such claim, damage, loss or expense (1) is attributable to bodily injury, sickness, disease or death, or to injury to or destruction of tangible property (other than the Work itself) including the loss of use resulting therefrom, and (2) is caused in whole or in part by any negligent act or omission of the Contractor, any Subcontractor, anyone directly or indirectly employed by any of them or anyone for whose acts any of them may be liable, regardless of whether or not it is caused in part by a party indemnified hereunder. Such obligation shall not be construed to negate, abridge, or otherwise reduce any other right or obligation of indemnity which would otherwise exist as to any party or person described in this Paragraph. In any and all claims against the Owner or the Owner's Representatives or any of their agents or employees by any employee of the Contractor, any Subcontractor, anyone directly or indirectly employed by any of them or anyone for whose acts any of them may be liable, the indemnification obligation under this Paragraph shall not be limited in any way by any limitation on the amount or type of damages, compensation or benefits payable by or for the Contractor or any Subcontractor under workers' or workmen's compensation acts, disability benefit acts or other employee benefit acts.

Contractor's liability insurance shall be purchased and maintained by the Contractor to protect him from claims under workers' or workmen's compensation acts and other employee benefit acts, claims for damage because of bodily injury, including death, and from claims for damages, other than to the Work itself, to property which may arise out of or result from the Contractor's operations under this Contract, whether such operations be by himself or by any Subcontractor or anyone directly or indirectly employed by any of them. This insurance shall be written for not less than any limits of liability specified in the Contract Documents, or required by law, whichever is the greater, and shall include contractual liability insurance applicable to the Contractor's obligations under Paragraph 6.4. Certificates of such insurance shall be filed with the Owner prior to the commencement of the Work.

The Contractor shall not employ any Subcontractor to whom the Owner's representatives or the Owner may have a reasonable objection. The Contractor shall not be required to contract with anyone to whom he has a reasonable objection.

CORRECTION OF WORK

The Contractor shall promptly correct any Work rejected by the Owner's representatives as defective or as failing to conform to the Contract Documents, whether observed before or after Substantial Completion and whether or not fabricated, installed or completed, and shall correct any Work found to be defective or nonconforming within a period of one year from the Date of Substantial Completion of the Contract or within such longer period of time as may be prescribed by law or by the terms of any applicable special warranty required by the Contract Documents. The provisions of this Article apply to work done by Subcontractors as well as to Work done by direct employees of the Contractor.

CHANGES IN THE WORK

The Owner, without invalidating the Contract, may order Changes in the Work consisting of additions, deletions, or modifications, the Contract Sum and the Contract Time being adjusted accordingly. All such changes in the Work shall be authorized by written Change Order signed by the Owner's representatives and the Contractor.

CONTRACT SUM/PROGRESS PAYMENTS

The Owner shall pay the Contractor for performance of the work, subject to additions and deductions by approved Change Orders, the Contract Sum of $____. The Contract Sum is determined as follows:

Base Bid ____________________________

Addenda ____________________________

Contract Sum ____________________________

Based upon invoices submitted to the Owner's Representatives, the Owner shall make payments on account of the Contract Sum to the Contractor as follows:

Draw 1	_____%	$__________
Draw 2	_____%	$__________
Draw 3	_____%	$__________
Draw 4	_____%	$__________
Draw 5	_____%	$__________

Payments may be withheld on account of

1. defective work not remedied,
2. claims filed,
3. failure of the Contractor to make payments properly to Subcontractors or for labor, materials, or equipment,
4. damage to the Owner or another contractor, or
5. persistent failure to carry out the Work in accordance with the Contract Documents.

Final payment shall not be due until the Contractor has delivered to the Owner a complete release of all liens arising out of this Contract or receipts in full covering all labor, materials and equipment for which a lien could be filed, or a bond satisfactory to the Owner indemnifying him against any lien. If any lien remains unsatisfied after all payments are made, the Contractor shall refund to the Owner all moneys the Owner may be compelled to pay in discharging such lien, including all costs and reasonable attorneys' fees. Owner may withhold a retainage of 20% of all invoiced charges if the Contractor fails to complete all contract items upon submission of final invoice.

This Agreement entered into as of the day and year first written above.

OWNER	CONTRACTOR
______________________________	______________________________
______________________________	______________________________

ADDENDUM 18: BID COMPARISON RESULTS

TO: ______________________ DATE: ______________________

FROM: ______________________ CLIENT #: ______________________

RE: ____

The following bids were received at ______________________ and opened on ______________________, at ______________________.

NAME OF BIDDER:	AMOUNT OF BID:
______________________	______________________
______________________	______________________
______________________	______________________
______________________	______________________
______________________	______________________
______________________	______________________

Persons Reviewing Bids:

______________________ ______________________

______________________ ______________________

______________________ ______________________

We have reviewed each bid listed above, and we recommend ______________ to perform the work on ______________________ property.

______________________ ______________________

(Homeowner's Signature) (Co-Homeowner's Signature)

______________________ ______________________

(Date) (Date)

If there are any questions about the recommendation process, please call ______________ of ______________________, at (   )   -   ______________.

ADDENDUM 19: NOTICE TO PROCEED

CONTRACTOR:______________________________ DATE:_______________________

ADDRESS ______________________________ PROPERTY

______________________________ ADDRESS: ____________________

______________________________ ____________________

ESCROW #:____________________

Dear ________________________,

Pursuant to the Homeowner/Contractor Agreement signed on ______________, you are hereby notified to proceed with the rehabilitation work on the property captioned above. Upon receipt of this notice, you are responsible for the rehabilitation work on said building under the terms and conditions of the Agreement, Specifications and Plans, if any. Please be prepared to begin construction within _________ days of receipt of this notice.

Please acknowledge acceptance of this notice in the space provided below.

____
Signature of Owner

____
Owner's Phone Number

Received the foregoing Notice to Proceed on _____________________
Date

Instructions to Contractor

Please sign both copies of this notice and return one copy to:

____
Contractor

____
Signature

____
Contractor's Phone No.

____
Rehab Specialist

____
Title

____
Rehab Specialist's Phone No.

ADDENDUM 20: EXPLANATION OF DRAW PROCEDURES

Dear Contractor:

Please be informed that progress draw inspections for Friday (same week) payments are performed on ______________________ (Date). Progress draw payments are made on Friday after 1:00 p.m. Inspections can be scheduled for Thursday and Friday, however, payment will not be made until the following Friday.

Effective immediately, contractors will need to submit a **Draw Inspection Request** form (enclosed) to ______________________ (Agency Name). These forms are to be submitted to the AGENCY Receptionist at least one day prior to the requested inspection day. Draw inspections will be scheduled on a first-requested, first-served basis.

Our inspectors are not expected to perform more than 6 progress or final draw inspections a day--combined. The time of day the inspection occurs shall be determined by the Inspector. If 6 inspections have already been scheduled for a day, the contractor will need to request that the draw inspection be performed another day or week.

Contractors should contact the Inspector between 8:00 a.m. and 9:00 a.m. Thursday morning to learn if they have completed enough work to receive a draw. Eligible contractors will need to stop by our offices on Thursday afternoon to obtain a list of documents (i.e. **notarized** lien releases, warranties, city inspection certificates) they will need to submit to ______________________ (Name) on Friday prior to payment.

Contractors will also need to obtain a **Disbursement Voucher** from us on Thursday afternoon. Contractors must secure the homeowner's signature on the voucher and return the voucher to ________________ (Name) on Friday prior to payment.

Having the homeowner sign the voucher will permit us to issue a check made directly payable to the contractor.

The earlier you request a draw inspection, the better your chances are for getting the inspection on the requested date. Please also be advised that contractors may be dropped a classification for prematurely requesting a draw.

Sincerely,

ADDENDUM 21: DRAW REQUEST

Borrower: ____________________

Address: ____________________

Contractor: ____________________

CONTRACTOR REQUEST FOR PARTIAL PAYMENT Draw #__________ $__________

I hereby certify that I have completed the work specified below in accordance with the Rehabilitation Contract and request partial payment of $__________ per the attached itemized request. I hereby certify that all payments due my subcontractors, suppliers and laborers have been made, and subcontractor and material house release of liens are attached.

____________________ ____________________

Authorized Agent for Contractor Date

BORROWER APPROVAL

I hereby accept all work completed by the contractor listed above and authorize partial payment of the amount specified in accordance with the above referenced Rehabilitation Contract, as due and payable to the contractor, from the proceeds of my loan.

____________________ ____________________

Borrower's Signature Date

REHAB SPECIALIST APPROVAL

Total Amount Approved: ____________________

Retained: __________10%__________

Net Amount Approved: ____________________

Progress Payment: ____________________

____

____

____

____

Inspector: ____________________

Title: ____________________

Date: ____________________

ADDENDUM 22: PARTIAL RELEASE OF LIENS

To All Whom It May Concern

In consideration of the sum of $1.00, or ________________________________ Dollars, and other valuable consideration in hand paid, the receipt whereof is hereby acknowledged, the undersigned does hereby waive, release, and relinquish any and all liens or claims, or right to lien or claim, for labor or materials, or both, furnished to date hereof, for premises known and described as:

Street
&
Number __________________________City________________________________

County __________________________State_______________________________

Witness the hand and Seal given this _______________day of

__________19____City of _____________________State of _______________

(Sign)________________________________
(Firm's Name)

(Trade________________________)

By__________________________________
(Authorized Agent)

ADDENDUM 23: CHANGE ORDER

DATE: ______________________________ CHANGE ORDER # _______________

JOB ADDRESS: ____

ORIGINAL CONTRACT AMOUNT: $_________ ADJUSTED CONTRACT AMOUNT: $________

ITEMS TO BE DELETED FROM SPECIFICATIONS

SPEC NO.	DESCRIPTION	AMT. DELETED
______________	____	________________
______________	____	________________
______________	____	________________
______________	____	________________
______________	____	________________
______________	____	________________
		________________
	TOTAL DELETIONS __________	$_______________

ITEMS TO BE ADDED TO SPECIFICATIONS

SPEC NO.	DESCRIPTION	AMT. ADDED
______________	____	________________
______________	____	________________
______________	____	________________
______________	____	________________
______________	____	________________
______________	____	________________
	TOTAL ADDITIONS	$_______________
	NET CONTRACT CHANGE	$_______________

EXAMINED AND APPROVED BY:

________________________________ ____________________________________
Rehabilitation Specialist Contractor

________________________________ ____________________________________
Homeowner Date

ADDENDUM 24: NOTICE OF FINAL INSPECTION AND FINAL PAYMENT APPROVAL

PROPERTY OWNER	CONTRACTOR
ADDRESS	ADDRESS
CASE NO.	PHONE NO.

This is a notice of final inspection of the repair work performed on your property. The contactor has indicated that all the work required by the work write-up and in the contract agreement has been satisfactorily completed. If you disagree with the contractor that all the work has been satisfactorily completed, now is the time for you to advise us and the contactor of any omissions, problems, or defects in the work. If you certify to us that the work is complete and in accordance with the construction contract, with no omissions, problems, or defects, we will issue a check payable to you and the contractor from the escrow account making final payment for the work. By signing this form, you are certifying that you have inspected the property and that the contractor's work has been completed in accordance with the construction contract.

Thank you for your cooperation in the repair of your property.

Sincerely,

Rehabilitation Supervisor

I acknowledge receipt of this notice of final inspection, and I certify that all work has been completed in accordance with the contract. In the event that it becomes necessary to rework or repair any of the listed items, I will contact the Contractor.

DATE	
WITNESS	PROPERTY OWNER

ADDENDUM 25: FINAL DRAW REQUEST

JOB ADDRESS: ____

CONTRACTOR: ____

CONTRACTOR REQUEST FOR FINAL PAYMENT $________________

I hereby certify that I have completed 100% of the work in accordance with the Rehabilitation Contract and request final payment of $________ which includes payment for all Change Orders of $________.

______________________________ ______________________

Authorized Agent for Contractor Date

FINAL PAYMENT CHECKLIST

Prior to receiving your final payment you must present to The Agency Rehab Specialist properly completed and notarized Lien Releases, Local County or City Inspection Certificates, and warranties.

	REQUIRED	RECEIVED
Contractor Affidavit	_____	_____
LIEN WAIVERS		
General Contractor	_____	_____
HVAC	_____	_____
Electrician	_____	_____
Plumbers	_____	_____
Other ________________........	_____	_____
LOCAL INSPECTION CERTIFICATES		
General Repair Permit...........	_____	_____
Plumbing Final..................	_____	_____
Electrical Final................	_____	_____
Heating Final...................	_____	_____

WARRANTIES	REQUIRED	RECEIVED
Roof	_____	_____
HVAC	_____	_____
Water Heater	_____	_____
Siding	_____	_____
Other ____________________	_____	_____

Rehab Specialist Approval:

I certify all work is complete; warranties and release of liens have been received; permits have been finalized, and I approve release of this final payment.

____
Rehab Specialist

____
Date

ADDENDUM 26: FINAL RELEASE OF LIENS

AFFIDAVIT OF (Sub Original) CONTRACTOR

....................Ohio,..........19...

STATE OF OHIO,.......................COUNTY, ss:

..............................being first duly sworn says that he is

..............................of.............................the

Sub/Original contractor having a contract with........................

the...

for...

situated on or around or in front of the following described property:...

....

whereof................................was the owner, part owner or lessee.

Affiant further says that the following shows the names of every subcontractor in the employ of said............................ .... giving the amount, if any, which is due, or to become due, to them, or any of them, for work done or machinery, material or fuel furnished to date hereof, under said contracts.

NOTE: This statement must be accompanied by a similar sworn statement signed by each of the subcontractors listed below.

SUBCONTRACTORS

NAME	TRADE	Amount due or to become due for work and material furnished to date hereof.

Said affiant further says that the following shows the names every person furnishing machinery, material or fuel, to...................... .... giving the amount if any, which is due, or to become due, to them or any of them, for machinery, material or fuel furnished to date hereof, under said contracts.

MATERIAL MEN

NAME	TRADE	Amount due or to become due for work and material furnished to date hereof.

Said affiant further says that the following shows the names of every unpaid laborer in the employ of ....furnishing labor under said contract, giving the amount, if any, which is due, or to become due for labor done to date hereof:

NOTE: If the fact is that every laborer has been paid in full, then recite: "Every laborer has been paid in full."
If not, then give each unpaid laborer's name and the amount due or to become due.

LABOR

NAME	HOURS	Amount due or to become due for labor furnished to date hereof.

That the amounts due or to become due to said subcontractors, material men and laborers, for work done, or machinery, material or fuel furnished to the date hereof, to... is fully and correctly set forth opposite their names, respectively, in the aforesaid statements, and further evidenced by certificates of every person furnishing machinery, material or fuel, hereto attached, and made a part hereof.

Affiant further says that .... has not employed or purchased or procured machinery, material or fuel from, or subcontracted with any person, firm or corporation, other than those above mentioned, and owes for no labor performed, or machinery, material or fuel furnished under said contracts, other than above set forth.

....

SWORN TO BEFORE ME AND SUBSCRIBED IN MY PRESENCE, AT................. Ohio, this ....day of .....................A.D., 19....

....
Notary Public

CERTIFICATE OF MATERIAL MEN

................Ohio, ...19...

The undersigned certify that to the date hereof they have furnished machinery, material or fuel as set out herein to ....for .... situated on or around or in front of the property described in the foregoing affidavit; that the nature of said machinery material or fuel furnished, the date when they commenced furnishing the same and the amount now due or owing to each of them, is correctly stated and set opposite their respective names or they have been paid in full, if so acknowledged hereon.

NAME	Machinery, materials and nature of the same	Commenced Furnishing	Amount due or to become due to date hereof.

ADDENDUM 27: TYPICAL PUNCH LIST ITEMS
(Check off missing items and note location)

ELECTRIC — Location(s) of List Items

__ All outlets and switches work.
__ Plates are straight and level.
__ All outlet plates installed tight to drywall.
__ Panel labeled for new circuits.
__ All lights have bulbs.
__ Final inspection completed.
__ Batteries in smoke detectors.
__ Other: ______________________________
__ Other: ______________________________

PLUMBING

__ All fixtures work; not chipped.
__ Fixtures are secure to wall or floor.
__ Air gap for dishwasher installed.
__ Tub is caulked.
__ Aerators/escutcheon plates installed.
__ All gas connections working.
__ Final permit inspection completed.
__ Other: ______________________________
__ Other: ______________________________

HEATING - COOLING

__ System responds to thermostat.
__ Warrantee and instructions provided.
__ Duct work free of debris.

DOORS, WINDOWS

__ All exterior doors weatherstriped.
__ Doors checked for tightness, lasp allignment and lock function.
__ Keys available.
__ Windows have locks that tightly close.
__ Window operates smoothly.
__ Screen in place.
__ Glass clean.

FINISHES

__ Paint removed from hardware, glass.

__ MISC.

__ Trash removed.
__ Other: ______________________________

ADDENDUM 28: SAMPLE CONTRACTOR'S WARRANTY

The undersigned hereby guarantees all of the work required and performed by the contract, one year from the date shown on the NOTICE OF COMPLETION as dated _______________, 19__. Contractor also attaches herewith all manufacturers' and suppliers' written guarantees and warranties covering material and equipment furnished under the contract.

IN WITNESS WHEREOF, the undersigned has signed and sealed this instrument this ________________ day of ____________________, 19 ____.

____

Contractor-Business Name

By: ________________________________

____

Witness Signature

ADDENDUM 29a: CONTRACTOR EVALUATION BY REHAB SPECIALIST

NUMBER		YES	NO
1.	Did the work progress without unreasonable or unexplained delays?	___	___
2.	Was there excessive inconvenience for the homeowner?	___	___
3.	Was the contractor available during the job?	___	___
4.	Was the contractor receptive to program concerns (e.g., program image, working relationship with homeowner, organization and procedures)?	___	___
5.	Was the contractor willing to adjust to job changes and problems?	___	___
6.	Did the contractor follow the specifications?	___	___
7.	Were technical guidelines (materials, quality, methods) followed?	___	___
8.	Were attempts made by the contractor to deviate from agreed-upon procedures or work?	___	___
9.	Did the contractor respond to problems and complaints adequately?	___	___
10.	Were appropriate behavior, attitude, and working relationships maintained by the contractor?	___	___
11.	Should this contractor be considered for additional jobs?	___	___

PUT REMARKS ON ANY ITEMS WITH NEGATIVE EVALUATIONS HERE: ______________________

____

____

____

____

____

____

____

ADDENDUM 29b: CONTRACTOR EVALUATION BY HOMEOWNER

NUMBER		YES	NO
1.	Did the work progress without numerous delays?	___	___
2.	Were you inconvenienced beyond what you were told to expect?	___	___
3.	Was the contractor available to discuss any questions you may have had?	___	___
4.	Were the workers courteous and professional?	___	___
5.	Did the contractor attempt to sell you additional work? Explain: ______________________________ ______________________________	___	___
6.	Did the contractor suggest changes as to how the work should be done? Explain: ______________________________ ______________________________	___	___
7.	Was the completed work what you expected?	___	___
8.	Would you recommend this contractor to another homeowner?	___	___
9.	Did the contractor respond to problems and complaints adequately?	___	___
10.	Did the contractor explain all procedures and warranties adequately?	___	___
11.	Please rate the quality of the work completed (circle one)		

Very Satisfactory	Satisfactory	Unsatisfactory	Very Unsatisfactory

EXPLAIN NEGATIVE EVALUATIONS BELOW: ______________________________

____

____

____

____

____

____

ADDENDUM 29c: PROCEDURES TO EVALUATE CONTRACTORS

The evaluation of contractors should be documented in writing, especially if the evaluation may be used in any future action against the contractor. Following are some suggested procedures to accomplish this.

1. Have homeowner complete an evaluation form or write a letter at the time of the final inspection.

2. Have the Rehab Specialist evaluate the contractor in writing upon completion of the work.

3. If there were problems during the warranty period, have the homeowner and the Rehab Specialist complete another evaluation form concerning the contractor's handling of the complaint.

4. Send follow-up evaluation forms to homeowners at a specified time (e.g., 6 months or a year) after completion of the work to assess continued satisfaction.

5. A qualified Rehabilitation Specialist should verify complaints by homeowners to determine the validity of any negative homeowner evaluation of the contractor's workmanship.

6. Keep written documentation of any complaints about the contractor made by subs or suppliers.

7. Keep records of failed code inspections and other evaluation by code inspectors.

8. Record any information relating to contractor performance that may be discovered during periodic HUD program reviews.

ADDENDUM 30: AUTOMATING SPECIFICATION WRITING AND ESTIMATING
FOR HOUSING REHABILITION

Automating Specification Writing and Estimating for Housing Rehabilitation

A Training and Technical Assistance Package including SPECMASTER® software

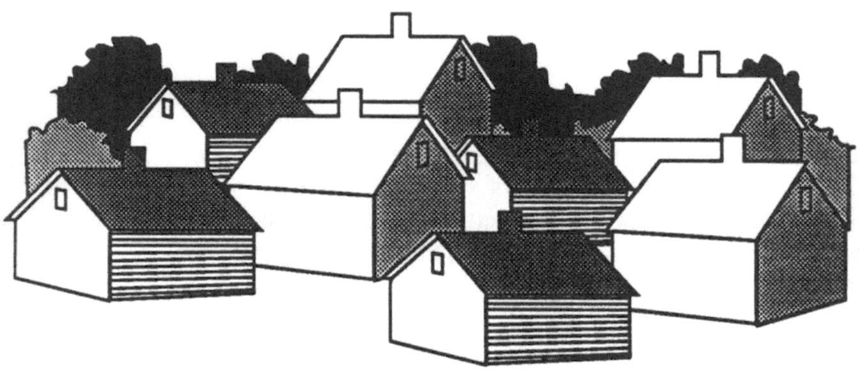

Offered by:

The Enterprise Foundation
500 American City Building
Columbia, Maryland 21044
(800) 624-4298

The Problem

As budgets are slashed and the need for affordable housing increases, today's city housing departments and non-profit housing development corporations are being asked to produce more output with less resources than ever before. Budgets to hire enough qualified housing specialists are so thin that most departments have a serious shortage of manpower.
Do the problems below sound familiar to you?

- Most or all of our work write-ups and estimating are done manually. *Would you benefit from automating your specs and estimates?*
- Our rehab specialists spend so much time on paper work that our field work is backlogged. *Would you benefit from having your rehab specialists freed from the tedium of paperwork?*
- There is a lack of consistency in our specifications, pricing, and work write-ups. *Would you benefit from a system that prepares reports for you consistently and automatically?*
- The number of housing units needing rehab is growing, but our ability to handle the need is not. *Would you benefit from your rehab specialists' ability to handle more cases in the same amount of time?*

The Solution

The Enterprise Foundation's *Training and Technical Assistance* package includes all the software, training, and technical assistance you need to fully automate your specification writing and estimating activities. Designed for organizations which spend much of their time working on selective rehab or home repair (rather than gut rehab or new construction), the *Training and Technical Assistance* package (using SPECMASTER® software) enables you and your organization to be more productive. You will produce more units per rehab specialist.

What is Specmaster?

SPECMASTER is a valuable, time-saving software tool that helps you produce work write-ups and cost estimates for single and multi-family housing rehabilitation. SPECMASTER has enjoyed widespread success with city-housing departments and nonprofit housing-development companies across the nation. SPECMASTER was conceived and developed by **The Enterprise Foundation's** *Rehab Work Group* (Columbia, MD) and programmed by **TSA/ADVET** (Pittsburgh, PA).

What's Involved in Automating?

Automating your specification writing and estimating system using SPECMASTER, often means changing of some your staff's responsibilities and procedures. Higher rehab volume is made possible by automation, but as your volume increases, the duties of your staff become more specialized.

Use an Entry Clerk

By using a data entry clerk to do all the inspectors' input, rehab specialists are able to concentrate on field work rather than office work.

Use an Inspection Checklist

A custom, inspector's checklist is used by the inspector as a recording sheet for work items, quantities and notes during the inspection. It then is used by the entry clerk to input the inspection into the computer.

Use a Database of Specifications and Prices

Automating your specification writing and estimating implies the need for standard pricing and specification language for the specs you use. SPECMASTER comes complete with a database of 1100 specifications and prices. It has the capacity to handle 10,000.

What Technical Assistance is Provided to Management?

Rehabilitation experts from **The Enterprise Foundation's** ***Rehab Work Group*** work with your management staff to evaluate the goals and needs of your department. We recommend ways to coordinate a smooth transition to a higher volume, more efficient housing-production system.

- We explain the cycle of activities through a SPECMASTER project.
- We create a customized inspector's checklist for you, and show you how to keep it up-to-date for the greatest efficiency.
- We customize your SPECMASTER database and explain how to keep it up-to-date. Prices and spec language must be checked and edited (as needed), to create a custom database that is appropriate for your organization.
- We show you the paper trail protocol between members of your rehab staff as well as which reports to use.
- We help you develop a streamlined approval and payment system for contractor draw requests.

What Training is Provided?

Training and Technical Assistance Package includes SPECMASTER hands-on training for members of your staff. We perform an on-site inspection with your rehab specialists using your new inspection checklist. Training is tailored to the needs of each staff member's responsibilities.

What Results Can be Expected?

Automating your specification writing and estimating with our *Training and Technical Assistance Package* and SPECMASTER software has several benefits.

- Use of your manpower is much more efficient.
- More accurate, detailed, and consistent specs and estimates are produced.
- Scope of work changes are easy to make. Even when several items are changed, new specs and estimates are produced bid-ready in minutes.
- Turn around time from inspection to bid documents can be short as half a day.

A Sample of Our Success...

Rehab Work Group has provided on-site technical assistance to more than 100 non-profit groups and government agencies. Although the results of automation will vary from organization to organization, the follow two examples are typical of the results that can be achieved.

- A SPECMASTER user in Baltimore specs out and estimates 70 rehabs of rental row houses each year while using only 30 working days of a rehab specialist.
- In Chattanooga, Tennessee, a user of SPECMASTER who specializes in the single-family rehab, averages 100 units a year per rehab specialist.

Specmaster Hardware Requirements

To operate Specmaster, you need the following computer hardware.

- An IBM or 100% compatible computer
- 640K RAM
- DOS 2.1 or higher
- One hard disk drive with 5MB available
- One floppy disk drive
- A monochrome or color monitor
- A Printer

Request a Free Needs Analysis

Rehab Work Group can provide you with a free analysis of your software and training needs, prior to purchasing *The Training and Technical Assistance Package*. While we are happy to discuss your needs at any time over the phone (1-800-624-4298), we can best analyze your needs if you fill the *Needs Analysis Questionnaire* which is attached to the *Scope of Work* and send it to us with samples of your current work write-ups.

Further Information

For more information about the *Training and Technical Assistance* package including SPECMASTER software, please refer to the *Scope of Work* or contact

Kent Buhl
Rehab Work Group Training Division
500 American City Building
Columbia, MD 21044
(800) 624-4298

Index

MIX
Papier aus verantwortungsvollen Quellen
Paper from responsible sources
FSC® C105338

If you have any concerns about our products,
you can contact us on
ProductSafety@springernature.com

In case Publisher is established outside the EU,
the EU authorized representative is:
Springer Nature Customer Service Center GmbH
Europaplatz 3, 69115 Heidelberg, Germany

Printed by Libri Plureos GmbH
in Hamburg, Germany